A Sophisticate's Primer of Relativity

A Sophisticate's Primer of Relativity

Second Edition

By

P. W. BRIDGMAN

Introduction by

ARTHUR I. MILLER

Wesleyan University Press

MIDDLETOWN, CONNECTICUT

First Edition, 1962
Second Revised Edition, 1983
First Wesleyan University Press Paperback Edition, 1983

LIBRARY OF CONGRESS CATALOGING IN PUBLICATION DATA

Bridgman, P. W. (Percy Williams, 1882–1961.
A sophisticate's primer of relativity.
Reprint. Originally published: Middletown, Conn.:
Wesleyan University Press, 1962.
Includes index.
1. Relativity (Physics) I. Title.
QC173.55.B74 1983 530.1'1 82-2068
ISBN 0-8195-6078-2 (pbk.) AACR2

Published by Wesleyan University Press
Middletown, Connecticut

Manufactured in the United States of America

Contents

Introduction

P. W. Bridgman and the Special Theory of Relativity

by ARTHUR I. MILLER

Percy W. Bridgman wrote *A Sophisticate's Primer* (1962) for the reader "who feels the need to stand back a little for a critical scrutiny of what he has really got" (*SP*, p. 3).* This was his personal trademark as physicist and philosopher and it is present everywhere in Bridgman's précis of special relativity theory.

The reissue of *Sophisticate's Primer* is particularly welcome for it exhibits a quest for clarity in foundations in the style of Ernst Mach and Henri Poincaré, in whose philo-

* N.B. Papers and books are here cited as follows. Bridgman (1927) refers to Bridgman's book *The Logic of Modern Physics* in the bibliography. Cross-references to *Sophisticate's Primer* are indicated with the code (*SP*, p. ⋯). Bridgman's extant manuscripts are on deposit at the Harvard University Archives, and here cited by date, for example (MS 2 August 1959, p. ⋯). Some of the manuscripts were written over a period of days, with sequential paging, with the actual dates indicated. Full citations are given in the bibliography. I am grateful to the Harvard University Archives for permission to quote from these materials, and to Bridgman's daughter, Mrs. Jane Koopman, for permitting me to quote from the draft manuscripts of *Sophisticate's Primer*, which will be deposited in the Harvard Archives.

sophical lineage Bridgman had placed himself in his first book on the philosophy of science, *The Logic of Modern Physics* (1927). Bridgman's profound and far-reaching analysis of space, time, and causality in *Sophisticate's Primer* is presented on a level that is basic enough to be appreciated by a wide spectrum of readers. It is a fine example of an analysis of foundations tempered by a philosophical view that continues to be relevant—the operational point of view. As the philosopher of science Adolf Grünbaum wrote twenty years ago in the Prologue to the first edition of *Sophisticate's Primer*, Bridgman's reflections on fundamental questions in this book "merit the serious attention of a wide public as the mature products of a long-standing intellectual concern."

Of Bridgman's many incisive analyses in *Sophisticate's Primer* I shall focus on the ones concerning the properties of light, clock synchronization, and the concept of the observer. Comparison of Bridgman's unpublished manuscripts on relativity from the period 1922–60, on deposit at the Harvard University Archives, with his published philosophical writings between 1924 and 1959 enables us to study his development of these notions into the form they took in *Sophisticate's Primer*.

As a mature philosopher-scientist, Percy Williams Bridgman (1954) recalled his reaction in 1914 to the state of electrodynamics and special relativity theory: "The underlying conceptual situation in the whole area seemed very obscure to me and caused me much intellectual distress, which I tried to alleviate as best as I could." Bridgman's urge to clean up these subjects to his own satisfaction later resulted in his proposing the operational view in *Logic of Modern Physics*. At the time of Bridgman's philosophical awakening in 1914, the 32-year-old Harvard

Assistant Professor of Physics already had a solid background in philosophy which he had begun to study seriously while a High School Senior in Newton, Massachusetts.**

At Harvard University, where he was awarded the B.A. in 1904 and the Ph.D. in 1908, he took courses with such Cambridge luminaries as Josiah Royce and George Santayana, in which he was almost certainly introduced to the pragmatism of William James. *The Logic of Modern Physics* was published in 1927 when Bridgman, then Hollis professor of mathematics and natural philosophy, was deeply involved in the research in high-pressure physics for which he was eventually awarded the Nobel Prize.

While the seeds of Bridgman's operational view can be traced to his "distress" of 1914 and then to his work of 1916–22 toward clarifying dimensional analysis, it was his deliberations on Albert Einstein's special theory of relativity that really served to crystallize the operational view for him. This work Bridgman began in earnest in 1922.

Early Assessment: 1922–27

> The Relativity Theory of Einstein is the result of, and is resulting in, an increased criticalness with regard to the fundamental concepts of physics. . . . The general goal of criticism should be to make impossible a repetition of the thing that Einstein has done; never again should a discovery of new experimental facts lead to a revision of physical concepts simply because the old concepts had been too naive. Our concepts and general scheme of interpretation should be so broad and so well considered that any new experimental

** For biographical material on Bridgman (1882–1961), see Kemble (1970).

> facts, not inconsistent with previous knowledge, may at once find a place waiting for them in our scheme. A program of consideration as broad as this demands a critical examination not only of the concepts of space and time, but of all other physical concepts in our armory. I intend in the following to wander over this whole broad field of criticism . . . no concept is to be admitted which does not bring with it its complex of operations; in fact, unless there is the complex of operations the concept has no meaning.
>
> P. W. Bridgman (MS September 1923)

As Bridgman wrote in his notes of September 1923, had the physicists of 1905 been more critical of fundamental concepts, Einstein's work would have been unnecessary.[1]

---

1. For an analysis of the state of physical theory from 1890 to 1911 as it affected the emergence and early reception of the special relativity theory, I refer the interested reader to Miller (1981), which contains extensive references to the secondary literature and a new translation of Einstein's 1905 special relativity paper entitled "On the Electrodynamics of Moving Bodies."

Some of the aspects of the 1905 relativity paper that attracted Bridgman follow. Einstein based his view of physics on two axioms. The first axiom is a version of the principle of relativity from mechanics, widened to include electromagnetic theory:

> The laws by which the states of physical systems undergo changes are independent of whether these changes of state are referred to one or the other of two coordinate systems moving relatively to each other in uniform translational motion.

Then, instead of attempting to explain the failure of ether-drift experiments, Einstein proposed a second axiom:

> Any ray of light moves in the 'resting' system with the definite velocity c, which is independent of whether the ray was emitted by a resting or by a moving body.

In Section 1 of the relativity paper Einstein developed the theory's

Bridgman intended this sharply-worded commentary—which objects to such analyses of time as Poincaré's (1898)—to serve as an introduction to the operational view he was developing in his 1927 book *The Logic of Modern Physics*. There, after repeating essentially the same message, and after paying due respects to Einstein's critical analysis[2]

---

key notion—the definition of simultaneity in a single inertial reference system. But first he defined an inertial reference system ("one in which Newton's equations hold"), and then he proposed an in-principle measurement operation for position relative to a reference system by means of rigid measuring rods. Einstein next stressed the importance of a clear understanding of time because, after all, the coordinates of a moving particle are functions of time. These steps enabled him to formulate a version of physical theory that did not contain such apparently unmeasurable quantities as the earth's velocity relative to the ether, which had been eliminated in ether-based theories of electromagnetism by postulation of compensating effects like the Lorentz contraction.

Einstein went on to define operationally the concept of an event as an occurrence at a particular position, measured relative to an inertial reference system and registered by a clock at that position.

2. Bridgman (1927) elaborated further on Einstein's critical analysis of space and time: "It is precisely here, in an improved understanding of our mental relations to nature, that the permanent contribution of relativity is to be found. We should now make it our business to understand so thoroughly the character of our permanent mental relations to nature that another change in attitude, such as that due to Einstein, shall be forever impossible." Bridgman (in 1959a) noted how his "attitude had changed most drastically" on this point: "To me now it seems incomprehensible that I should ever have thought it within my powers, or within the powers of the human race for that matter, to analyze so thoroughly the functioning of our thinking apparatus that I could confidently expect to exhaust the subject and eliminate the possibility of a bright new idea against which I would be defenseless." Although in *Sophisticate's Primer* (*SP*, p. 4) Bridgman repeated in part his early assessment of Einstein's work on special relativity theory, his later statements (*SP*, pp.

at appropriate junctures, Bridgman wrote: "In general, we mean by any concept nothing more than a set of operations; *the concept is synonymous with the corresponding set of operations*" (italics in original). Although the notion of defining a physical quantity operationally had been discussed even before Einstein, by Mach (1960) and Poincaré (1902), for example, it was clearly Einstein's use of this notion in the relativity paper that most impressed Bridgman. In fact, in *The Logic of Modern Physics* Bridgman entitled the section in which he introduced the operational point of view, "Einstein's Contribution in Changing Our Attitude toward Concepts."

At first, however, Bridgman was equivocal on the validity of the special relativity theory. This attitude is evident throughout *The Logic of Modern Physics.* He expressed it forcefully on 27 December 1922 in a lecture to the AAAS in Boston, in which he emphasized as well the theory's lasting philosophical importance (MS 27 December 1922):

> The physicist may therefore well doubt whether the theory of Einstein in its present form will ultimately survive, but entirely apart from the ultimate truth of the theory, there can be no question that matters can never return to their condition before the formulation of the theory, and besides addition to our knowledge of facts, Einstein has made changes in our points of view which must have their permanent effect. It is of these changes of view with respect to space and time that I wish to speak. Any one familiar with speculation before and after the theory cannot fail to remark the increase in self consciousness and self criticism of our

---

157, 160) have the effect of further removing the aura of logical positivism from the operational view.

> attitude toward measurements in general. Now what appears to be E's most important service, and the one whose effect is most likely to survive, is one which in its statement appears almost trivial, but which nevertheless had far-reaching effects. This is merely the insistence on the requirement that the quantities which we use in our equations (or concepts in general) have a meaning.

One point of concern to Bridgman was that there were no exact inertial systems in nature[3] and that, consequently, the "special theory of relativity may very probably be only a close approximation" (Bridgman, 1924). In addition, as he wrote in *Logic of Modern Physics*, there was the possibility that "new kinds of experience" would demonstrate the lack of exact validity of special relativity.

Besides the "rapidly increasing array of cold experimental facts," several matters of principle concerned Bridgman. First, although light was central to special relativity theory, the theory did not get to the heart of the problem of the nature and constitution of light; and this problem was exacerbated at the time by complexities due to developments in the quantum theory of radiation. Within the operational framework Bridgman (1927) viewed the problem of the nature of light as follows. Can we really consider light to be a "thing that travels, thing not necessarily connoting material thing"? Bridgman could attribute no operational definition to the notion that light exists at every position intermediate between source and sink because, from the "view of operations, light means nothing more

---

3. Einstein (1923) considered the concept of the inertial reference system to be a "logical weakness" of the special relativity theory. On the other hand, the 1915 general relativity theory included observers in accelerated reference systems.

than *things lighted* [and] light as a thing traveling must be recognized to be pure invention" based on a combination of sense perceptions and "ordinary mechanical experience." One could suppose, Bridgman continued, that the propagation of a light wave could be detected by a suitably placed series of screens, with each screen so constructed that it destroyed only an infinitesimal part of the beam. This arrangement "evaporates," however, because contemporaneous quantum theory conceives light not as an irreducible quantity, but as an "exceedingly complicated thing." For example, a much-debated conundrum of the 1920s was how light quanta could explain interference.*

Nor does it help matters to consider light in transit as a transfer of energy. Conservation of energy entails time, because one must "integrate over all space the local energy at a definite instant of time." But in order to spread the notion of time over space, special relativity assumes light to be a thing traveling; consequently, a vicious logical circle ensues to which Bridgman offers no resolution.

Since Einstein had, for "convenience and simplicity" in dealing with optical problems, chosen to consider light to be a thing traveling, he could use the customary concept of velocity, in which an observer at rest in a reference system determines the distance covered by a material thing in a measured time interval on distance and time scales at rest along the system's coordinate axes. But in special relativity theory the notion of light propagating led to situations that were not "at all satisfying logically": light is neither a wave

* Miller, (1978) Visualization Lost and Regained: The Genesis of the Quantum Theory in the Period 1913–1927, and (1982) Redefining *Anschaulichkeit*.

disturbance in a medium, for the medium is not observed to have any effect on the light, nor is light a material "projectile," for then its velocity would not be independent of its source's motion, as in Walther Ritz's theory.

Renouncing the concept of light as a thing traveling, continued Bridgman, enables one to use an alternative notion of velocity, namely, a self-measured velocity which is a "hybrid sort of thing" in that it combines the time on a moving observer's clock with distances marked off along a resting coordinate system. An example would be the velocity determined by an automobile rider through comparison of his clock with markers on the road.[4] Despite the peculiarity of mixing coordinates, the self-measured velocity is independent of how time is spread through space because it is a single-clock velocity (see also *SP*, p. 106). From the Lorentz transformations it follows that the self-measured velocity of light is infinite, and so it is not a physical velocity. Bridgman found nothing amiss in this result because it agreed with assuming that light is not a thing traveling and hence need no longer be thought of as having the customary "physical velocity."[5] He proposed no alternative theory of space and time to include an infinite self-measured velocity of light. But he emphasized that this tack could lead to a theory that would be independent of the structure of light, though the "simplicity and mathematical tractability"

4. In the (1927) book Bridgman did not use the term "self-measured" velocity (see *SP*, p. 26).

5. The self-measured velocity of light is what Bridgman referred to in *Sophisticate's Primer* as a "here and now" velocity, thereby eliminating any notion of light as a thing traveling. However, in *Sophisticate's Primer* Bridgman did not consider the self-measured velocity of light be to infinite (*SP*, p. 138).

indigenous to special relativity theory would very likely be lost. Bridgman considered the loss of these qualtities to be worthwhile if it led to a theory closer to the requirements of the operational view. In fact, he continued, the "whole problem of the nature of light is now giving the most acute difficulty."

Bridgman's conclusion was that whereas Einstein's special relativity theory succeeded in dealing with optical phenomena with a "simple mathematical formula . . . the explanatory aspect is completely absent from Einstein's work." But analysis of special relativity theory and Einstein's own description of it reveal that the theory makes no assumptions about the constitution of matter or light and attempts only to account for phenomena, i.e., it is a theory of principle. Furthermore, special relativity theory can discuss the propagational aspects of light quanta. Bridgman was dissatisfied with that sort of theory primarily because of difficulties concerning the nature and constitution of light. In his opinion "it would seem that we ought at least to start over again from the beginning and devise concepts for the treatment of optical phenomena which come closer to physical reality."

Toward Sophisticate's Primer: 1935–60

Comparing special relativity theory to general relativity theory in 1936, Bridgman emphasized the special theory's firmer experimental basis and its "fewer philosophical considerations"—that is, considerations of a nonoperational sort. As is discussed later, Bridgman never accepted general relativity theory. Yet he continued to be vexed by the problem of describing the propagation of light. He wrote (in

1936) that the "fundamental arguments of the special theory demand a light signal, thought of as expanding spherically through space and apprehended by many observers, and therefore consisting of many photons." Thus he accepted the exact validity of special relativity theory only for "comparatively large scale phenomena."

Bridgman thought highly of the new quantum mechanics because of what he took to be its firm operational basis.[6] On the other hand, he considered contemporaneous problems of relativistic quantum mechanics to be strong indicators of the inapplicability of special relativity theory to the microscopic domain. He suggested that nonrelativistic quantum mechanics offered a clue toward resolving the problem of the nature of light propagation because it showed it "unprofitable to attempt to visualize" such phenomena as the "photon as a thing that travels . . . we seem to be at the point where we must learn to get rid of this adventitious aid in thinking."

Over twenty years later Bridgman's attitude toward special relativity theory had hardly changed, despite breathtaking advances in quantum electrodynamics that had resulted in a consistent blending of special relativity theory and quantum theory. Thus in *Sophisticate's Primer* he wrote that "conventional relativity theory is not concerned with the second step toward the microscopic—that is, it does not concern itself with photons" (*SP*, p. 139, see also pp. 162–163). Almost certainly Bridgman's 1961 position on applying special relativity theory to the microscopic domain was the one he held in 1927, when he wrote that on the

6. Bridgman (1927) applauded advances in quantum theory recently achieved by "Heisenberg-Born and Schrödinger," who had emphasized understanding concepts in terms of "physical operations."

atomic side one could hardly expect that the concepts of space and time would "have the same operational significance . . . that they have on the ordinary scale." In the strict operational sense he was right, because in relativistic quantum mechanics special relativity is applied formally as Lorentz covariance; one does not speak of observers with inertial reference systems equipped with clocks and rods who ride on submicroscopic elementary particles.

Thus in *Sophisticate's Primer*, Bridgman accepted special relativity theory to be valid *ab initio* in the macroscopic domain and he wrote:

> The "critical" attitude of this essay is critical only to the extent that it is necessary to find out what the theory essentially is. The theory as ordinarily understood is accepted without question as a working tool which, in its ostensible universe of discourse, gives the best control we have yet been able to acquire of phenomena [although this] does not mean that all problems of understanding have been solved. This essay is directed toward these problems of understanding (*SP*, pp. 4–5).

The interpretation of relativity theory as a "control" over nature is consistent with Bridgman's operational point of view.

A set of notes written in 1959 pinpoints two principal problems with special relativity theory that concerned Bridgman:

> The conventional formulations of relativity seem to me unsatisfactory in at least two respects. In the first place, the device of using two observers in two different frames of reference, and the talk about how a certain phenomenon would appear to the moving observer and what he would

> say about it seems clumsy and raises the implication that the moving observer may be deceiving himself with respect to the "reality." Furthermore it does not correspond to what actually happens. I cannot get away from myself — what happens in the other frames of reference eventually has to be described and find its meaning in terms of what happens to me. I should not have to depend on what the moving observer tells me, but I, with my instruments, should be able to make the readings that correspond to what the conventional moving observer is supposed to do. I am to be the central clearing house, and all instrumental readings should be made in such a way as to be utilizable by the central clearing house. This does not obscure the fact, however, that I, the central observer, have to concern myself with moving sticks and moving clocks. In the second place, the conventional treatment of light is as a thing travelling — a light signal is thought of as an identifiable thing travelling which can be watched as it travels just as a moving wave on water can be watched. But light is not a *thing* travelling, and it should be handled in a way not to involve any unconscious consequences of thinking of it in this way. In order to do this, the whole question of the true nature of radiation, which has been bothering me for so long, will have to be considered (italics in original, MS 11 July 1959).

In the course of completing *Sophisticate's Primer*, Bridgman realized a means to respond operationally to the problem of interpreting how light propagates through space; this response also constituted part of his resolution to the first problem—namely, the reformulation of physics in terms of a single observer. Before we reach that point, let us accept for the moment Bridgman's conviction that there is "something progressive about light" (*SP*, p. 41), and turn to the nature of the velocity of light, and in particular whether it is isotropic and whether the one-way velocity of light can be measured. In order to set the stage for Bridgman's analysis let us digress briefly to survey previous work by Poincaré and Einstein.

In a remarkable paper of 1898 entitled, "La mesure du temps," Poincaré investigated the connections between the nature of the velocity of light and the notion of time. Poincaré demonstrated that astronomers measured the velocity of light received from stars by agreeing *ab initio* that space is isotropic for light propagation and that the velocity of light is the same in every direction. Although this assumption could not be tested experimentally, it conformed with the apparent symmetry of space, i.e., the "principle of sufficient reason," which in turn furnished support for it.[7] Consequently, astronomers agree also on the meaningfulness of an unambiguous value for the one-way velocity of light.

Poincaré's (1898) analysis of the epistemological problems inherent in defining a one-way light velocity was based on a variant of Hippolyte Fizeau's well known 1849 measurement of the velocity of light, which he accomplished as follows (Fig. 1): Let a ray of light be emitted from a point A at the time t_A as registered on a clock at rest at A; after traveling a known distance $\overline{AB}$ the light ray is reflected from a mirror at B and arrives back at A at time t'_A. Since only one clock is involved here no problems are created by the relation between phenomena that occur at widely separated

7. Poincaré's (1898) analysis was based on his view that the foundations of geometry allowed for no preferred directions in the space described by Newton's mechanics. For further discussion, see Miller (1981; in press).

spatial points, i.e., distant simultaneity. Thus, Fizeau measured the velocity of light to be

$$c = \frac{2\overline{AB}}{t'_A - t_A}. \tag{1}$$

Then he generalized this result to apply to the one-way light velocity from A to B and from B to A. To make this generalization, Fizeau had assumed implicitly the equality of the to-and-fro velocities and the isotropy of space for the propagation of light. In fact, he had measured only the to-and-fro velocity and then generalized this measurement to the equality of the velocities of light from A to B (c_{AB}) and from B to A (c_{BA}), i.e., $c_{AB} = c_{BA}$.

Poincaré addressed the question of whether c_{AB} or c_{BA} could be measured separately. For example, he wrote in "Measure of Time," suppose a clock were placed at B. Then the velocity of light c_{AB} is (Fig. 2)

$$c_{AB} = \frac{\overline{AB}}{t_B - t_A}. \tag{2}$$

A vicious circle results since the measurement of $(t_B - t_A)$ requires c_{AB}, and therefore "such a velocity could not be measured without measuring a time." As a result of this analysis he concluded that, to ensure the descriptive simplicity of Newton's laws of mechanics, he had to assume that the velocity of light was the same in every direction and that the one-way value was Fizeau's; and the proper definition of time is the one from Newtonian mechanics and is independent of the reference system's motion, in agreement with our sense perceptions.

Portions of Poincaré's 1898 analysis were rediscovered

by Einstein in 1905, who pushed them to startling conclusions that could be plumbed only within a philosophical framework that was to a large extent liberated from sense perceptions. After proposing in-principle operational definitions for distance and time, Einstein (1905) discussed a method for synchronizing clocks based on Fizeau's procedure for measuring the velocity of light that is a Gordian resolution of the situations in Figures 1 and 2 (Fig. 3). Einstein placed a clock at B, and then defined a "common 'time' for A and B" through the equation

$$t_B - t_A = t'_A - t_B \tag{3}$$

which means that

$$c_{AB} = c_{BA}. \tag{4}$$

But Einstein did not need to consider Eq. (3) or (4) a definition, since the equality of the to-and-fro velocities of light follows from the two axioms of relativity, which in turn imply the isotropy and homogeneity of space for light propagation. Without the two axioms of relativity, Einstein's placing of clocks at A and B would have removed neither the vicious logical circle exposed by Poincaré nor the necessity for Fizeau's implicit assumptions. Having broken the vicious logical circle of clock synchronization, Einstein went on to define the one-way velocity of light as $c_{AB} = c_{BA} = c$, where

$$c = \frac{\overline{AB}}{t_B - t_A}. \tag{5}$$

Bridgman's unpublished notes of 29 July 1959 reveal that he had devoted much effort to the design of an experiment that could yield the one-way light velocity, and he thought he had succeeded. Bridgman's analysis of the experiment in Figure 4 shows that an anisotropy is expected, although it would probably be extremely small. But, Bridgman continued, he would not be "scandalized by this result," because he could interpret the anisotropy as an indicator of how much the laboratory system differs from a true inertial system.

The problem of spreading time through space is linked with simultaneity and the definition of velocity. Bridgman's notes during 1959–60 reveal that he was much occupied with this problem, and he gave it extended treatment in *Sophisticate's Primer*. His interest in synchronizing clocks by methods that differed from Einstein's is expressed in some of his early unpublished notes. For example, in September 1923 he wrote that, in order to measure the to or fro velocity of light without becoming enmeshed in a vicious logical circle, "we must set our clocks at the two stations by other means than the optical means employed by Einstein" (p. 40).

Herbert E. Ives's series of papers during 1937–51 may well have been the catalyst for Bridgman's reconsidering this issue. Although Ives was a life-long critic of relativity, his criticisms were constructive and were offered at the highest levels of scholarship.[8] Since only the to-and-fro velocity of light has operational meaning, Ives considered that Einstein had no grounds for asserting that the to-and-fro velocities were equal to the round-trip velocity. Ives's basic objection was to the principle of the constancy of the velocity of light, which he found "not only 'ununder-

8. In a classic experiment of 1938 Ives and G. R. Stilwell confirmed empirically the phenomenon of time dilation, although they considered that their data supported the Larmor-Lorentz ether-based theory (see Miller, 1981).

standable,' [but] *not* supported by 'objective matters of fact' " (Ives, 1951). In order to avoid using light signals to synchronize clocks, Ives suggested an alternative method of infinitely slow clock transport which depended on the self-measured velocity of the transported clock (see *SP*, pp. 65–66). He succeeded in deducing somewhat complicated Lorentz-type transformations that were functions of the self-measured velocities of moving rods and clocks, but which reduced to the customary Lorentz transformations when those velocities became zero. Ives (1951) said that his Lorentz-type transformations were derived by the "operational principle championed by P. W. Bridgman." This comment evidently prompted Bridgman to undertake a detailed critique of Ives's procedures (MS 18 August 1959, p. 3). Although Bridgman was not in total agreement with Ives's derivation of the Lorentz-type transformations and his applications of the operational method, he became interested in the notion of infinitely slow clock transport (*SP*, pp. 64–66). In view of Bridgman's definition of operational meaning from the 1927 *Logic of Modern Physics*, it may seem surprising to find him in 1961 advocating a version of infinitely slow clock transport in which the self-measured velocity of a transported clock is extrapolated to zero. Ives did not go to that extreme but wrote: "[Permitting the self-measured velocities to go to zero] means that the setting clocks would not be moved, the distant clocks would have no epoch allocated, and the measurement could not be made" (Ives, 1951). Bridgman criticized this assertion in *Sophisticate's Primer* (*SP*, p. 66), and in notes written in 1959 in a manner that illuminates how he had further developed the operational view since 1927: "It seems to me [that Ives's] argument is fallacious. Operational methodology does not mean that the limiting process be carried out *physically*—there is no reason why the physical operation should not be carried out for several values of [the self-measured velocity] different from zero and then mathematical extrapolation be made to zero" (italics in original, MS 18 August 1959).[9] For in Bridgman's developed form of operationalism, "mathematical extrapolation" is a well-defined mental activity. Bridgman's *ab initio* acceptance of special relativity theory in *Sophisticate's Primer* led him to conclude that clocks set by infinitely slow clock transport would agree with those set by Einstein's method (*SP*, p. 66).

The Conventionality of Distant Simultaneity

Einstein's method of synchronizing clocks using light signals can be reformulated so as to emphasize the conventional, i.e., the nonempirical, status of the one-way velocity of light. This method of analyzing simultaneity had been investigated systematically by Hans Reichenbach (1958) and later elaborated on by Adolf Grünbaum (1973). In his notes and in *Sophisticate's Primer*, Bridgman devoted considerable thought to the so-called Reichenbach-Grünbaum thesis of the conventionality of distant simultaneity.

Reichenbach generalized Einstein's Eq. (3) to

$$t_B = t_A + \epsilon\,(t'_A - t_A), \tag{6}$$

where in agreement with the principle of causality $0 < \epsilon < 1$. The choice of $\epsilon = 1/2$ corresponds to Einstein's special relativity theory in which the space of inertial ref-

9. Bridgman had further developed his notion of operational definition in several works (e.g., 1938, 1959a), after the 1927 publication.

erence systems is isotropic and homogeneous for the propagation of light, in which the definition of clock synchronization is transitive and symmetric, and where the symmetrical Lorentz transformations hold. Other values of ϵ correspond to $c_{AB} \neq c_{BA}$ because from Eq. (5) it follows that

$$c_{AB} = c/2\epsilon \tag{7}$$

and

$$c_{BA} = c/2(1 - \epsilon); \tag{8}$$

as required by the operational meaning of the round-trip light velocity

$$\frac{1}{c_{AB}} + \frac{1}{c_{BA}} = \frac{2}{c}. \tag{9}$$

According to Reichenbach, Einstein chose $\epsilon = ½$ for descriptive simplicity. Since any value of ϵ may be chosen within the open interval $0 < \epsilon < 1$, then certain predictions of a special relativity are dependent on the choice of ϵ—that is, are conventional. Bridgman set out to investigate the conventional aspects of special relativity in a manner that went beyond anything hitherto written on this subject. In the MS of 27 July 1959, pp. 3–4, Bridgman wrote:

> Given now a "stationary" single-clock system, we shall of course find that uniformly moving meter sticks are shortened and the "rates" of uniformly moving clocks are altered. A first question to be answered is whether the shortening and the change of rate are given by the Lorentz formulas for inertial systems? One would suspect yes. Let us examine this in detail.
>
> Imagine two "stationary" systems (1) and (2) with identical geometrical mesh systems, and with locally distributed clocks running at the same rate, but with different zero settings $x_1 = x_2$.[*]

Set the clocks by despatching a light signal from the origin and reflecting it back from a mirror at the point x, the clock at the origin in either (1) or (2) reading $2x/c$ when it returns. Assume the Einstein method of setting clocks in (1) and assume that in (2) the settings are consistent with a real difference of "go" and "come" velocity of light and that the clocks give this real difference.

Time of arrival of signal in (1) is x/c

" " " " " " (2) satisfies $t_2 = \epsilon(2x/c)$[†]

(Ives, Grünbaum)

$$\text{Hence } t_2 - t_1 = \frac{x}{c}(2\epsilon - 1) = \alpha x/c,$$

where $0<\alpha<1$. This holds at all times. The difference of clock settings entails a difference of velocities measured in the two systems.

Imagine an object moving with velocity v_1 in (1) leaving the origin at $t_1 = t_2 = 0$, and arriving at $x = 1$ at t_1. Then

$$v_1 t_1 = 1$$

The same object in (2) satisfies

$$v_2 t_2 = 1$$

$$t_2 - t_1 = \alpha x/c = 1/v_2 - 1/v_1 \qquad 1/v_2 = 1/v_1 + \alpha x/c$$

Hence

$$v_2 = \frac{v_1}{1 + \alpha v_1/c} \qquad v_2 = \frac{v_1 c}{c + \alpha x v_1} : x = 1$$

Hence velocities differ in (1) and (2) by quantities of the first order.

What now about the shortening of meter sticks? We are concerned with *two* moving frames — (1) with its system of clock setting and (2) with its. In the (1) system we assume that the Lorentz transformation holds

$$\left.\begin{aligned} x'_1 &= \frac{x_1 - v_1 t_1}{\sqrt{1 - \beta_1^2}} \\ t'_1 &= \frac{t_1 - \frac{v_1}{c^2} x_1}{\sqrt{1 - \beta_1^2}} \end{aligned}\right\}$$

The contraction of the moving meter stick in (1) is the value of x_1 when $x'_1 = 1$ and $t_1 = 0$: $x_1 = \sqrt{1 - \beta_1^2}\, x'_1$, the conventional value. What now is the contraction in (2): This is the value of x_2 at $t_2 = 0$ corresponding to $x' = 1$. When $t_1 = 0$ at $x_1 = \sqrt{1 - \beta_1^2}$, $t_2 = ax/c = \frac{a}{c}\sqrt{1 - \beta_1^2}$. The end of the meter stick has overshot its mark. Where was it at the time $\frac{a}{c}\sqrt{1 - \beta_1^2}$ earlier? It has been moving with velocity v_2, hence the distance $a\frac{v_2}{c}\sqrt{1 - \beta_1^2}$ and x_2 when $t_2 = 0$ is $\sqrt{1 - \beta_1^2}\left(1 - \frac{av_2}{c}\right)$.

The important question now is whether the Lorentz transformation applies to (2). In other words is

$$\sqrt{1 - \beta_1^2}\left(1 - \frac{a\, v_2}{c}\right) = \sqrt{1 - \beta_2^2} = \sqrt{1 - \left(\frac{v_2}{c}\right)^2}$$

or

$$\sqrt{1 - \left(\frac{v_1}{c}\right)^2}\left(\frac{1}{1 + \frac{a\, v_1}{c}}\right) = \sqrt{1 - \left(\frac{v_1}{c}\right)^2\left(\frac{1}{1 + \frac{a\, v_1}{c}}\right)^2}$$

evidently not for

$$\sqrt{1 - \left(\frac{v_1}{c}\right)^2} \neq \sqrt{\left(1 + \frac{av_1}{c}\right)^2 - \left(\frac{v_1}{c}\right)^2}.$$

---

[*] Since Bridgman assumed "identical geometrical mesh systems," the coordinate grids in (1) and (2) are identical.

[†] The result for t_2 follows from Eq.(7).

Thus Bridgman found that the root of the conventionality of distant simultaneity resides in the difference in settings for the zero points (i.e., of synchronization) between two clocks running at the same rates in their rest systems. Furthermore, a consequence of this difference in synchronization is a conventionality of relative velocities, which in turn leads to the result that the velocity of (1) relative to (2) is not the negative of the velocity of (2) relative to (1); basic symmetries of physics are dropping by the wayside.

Bridgman's conclusion on his final equation is:

> This result is at first disconcerting. The shortening of the meter stick is not given by the same formula when the method of setting clocks is changed. Does this mean that the "shortening of the meter stick" contains an element of convention in the canonical transformation? This does not feel right. Physically it looks as though it ought to be possible to give a meaning to "length of a moving meter stick" irrespective of the method of setting clocks (MS 27 July 1959, p. 4).[10, 11]

10. In his notes (MS 1 August 1959, p. 10), Bridgman recalled hearing about a method proposed by Einstein to determine length that involved two meter sticks of equal rest lengths moving in opposite directions with the same velocity, a method that was independent of any element of conventionality. In *SP* (pp. 92–93) Bridgman elaborated on that method. His critique is incorrect, however, because the velocities need not be equal and calculation shows that the length of the moving rods measured in the stationary system is less than either of their rest lengths. Einstein (1911) proposed this method without offering any calculation. See Winnie (1972) for details. Among other cases that are independent of conventionality are the round-trip time dilation effect and the increased distance of travel for unstable elementary particles that are traveling close to the velocity of light (see Winnie, 1970).

11. In MS 26 May 1960 (pp. 11–12) Bridgman speculated on the possibility of framing physical theory in a nonconventional mathematics

A few pages later Bridgman wrote:

> There is something funny about the "conventionality" of the single directional velocity of light. Now different values of ϵ in the stationary system correspond to different methods of setting clocks. Each one of these methods should give a physically consistent description of all physical phenomena if the precise choice of ϵ is a convention, and therefore to each different stationary system with a different ϵ, there should correspond a Lorentz transformation. Perhaps the answer is that the Lorentz transformation is not the most general one satisfying the requirements of relativity. [Bridg-
> [my][*]
> man inserted this sentence on 10 August 1959.] But all our

---

that conformed to the operationalism view and so avoided from the start the notion of one-way light velocity. He went on to write:

> No sooner is a system of coordinates established in which the time variable is spread through space than we *have* to talk about "one-way" velocity, willy nilly, for the time derivative of the space coordinate is such a one-way velocity.
>
> The considerations of the last paragraph offer, I believe, a logical way out of the dilemma in which we apparently find ourselves. I believe, however, that average physicists, among whom I count myself, will react in a somewhat different way. The physicist has found conventional mathematics applicable in such an overwhelming number of physical situations, that I think he will react to the analysis above by the expectation that if the propagation of light is not isotropic, the failure of isotropy will be found to be very small. I think he would be very much surprised if future experiment ever showed that under some conditions a value for ϵ differing from ½ by first order terms might be demanded by experiments in terrestrial laboratories, a possibility which the analysis of Reichenbach leaves open (italics in original).

Bridgman provided no details for these private speculations, and he crossed the pages with diagonal pencil marks. By April 1961, he had concluded that the one-way light velocity is unmeasurable (see pp. xxxii–xxiv below).

[*] The [my] is Bridgman's own insertion.

> analysis seems to show that this is not possible (MS 1 August 1959, p. 10).

At least one aspect of the conventionality thesis that he may have considered "funny" he went on to describe thus:

> If it is only the go-come velocity of light that has operational meaning there is going to be an important reaction on conventional arguments in cosmology, the light from distant parts of the universe being unidirectional (ibid.).

Poincaré had in 1898 also stressed the point that it would be impossible to do research in astronomy without assuming the isotropy and homogeneity of space for light propagation, and that the one-way velocity of light was c.[12]

Detailed investigation of the conventionality of length and time in which ϵ is different in different inertial systems entails deducing ϵ-dependent Lorentz-type transformations, which Bridgman's unpublished notes reveal he was unable to derive. Yet he was convinced that the Lorentz transformation was the most general transformation satisfying special relativity. Bridgman's placement of the insert into the passage of 1 August 1959, quoted from above, emphasizes that he posed and then found to be unworkable the notion that the "Lorentz transformation is not the most general one satisfying the requirements of relativity." Thus he was led to conclude that the value of ϵ had to be the same in every inertial reference system, and since special relativity was valid, then ϵ had to be ½. This was pure

---

12. But Poincaré's conclusion held strictly only for astronomy. For in ether-based electromagnetic theory the space of inertial reference systems was in-principle not isotropic and homogeneous for the propagation of light. However, the null results of ether-drift experiments asserted otherwise (see Miller, 1981, esp. Chapter 1).

speculation, however, since he lacked the mathematical proof.

In September of 1960 Bridgman returned to the experiment in Figure 4, and found a formula for the anisotropy of light which was consistent with no difference of measured transit times from A to B and from B to A.[13] This result was perplexing because, as he had written in the manuscript of 30 July 1959 (see Fig. 4): "If these two Δ's are the same then the difference between the go and come times is not measurable, and their equality is a convention." Thus, he had to conclude in 1960 (MS 4 September 1960):

> It would seem that we are here confronted with a genuine logical and semantical dilemma. Is there any meaning to the [v_{AB} and v_{BA}] which we have used in our analysis beyond a purely paper-and-pencil meaning? Furthermore, if the principle of relativity is true, then the propagation of light must be isotropic. But there may be a conspiracy of nature which prevents us from checking whether light propagation is isotropic, and therefore, in so far, prevents us from checking whether the principle of relativity is true. In other words, there may be a conspiracy of nature to prevent us from checking the truth of the principle of relativity. If this is the case, then the principle of relativity itself must be recognized to be a "convention."

But earlier in this manuscript he had developed the proof that would appear in the published version of *Sophisticate's*

13. He wrote (MS 4 September 1960, p. 8):

$$\frac{1}{v_\theta} = \frac{1}{c} + \frac{1}{2}\left(\frac{1}{v_{AO}} - \frac{1}{v_{BO}}\right)\cos\theta$$

where the angle θ is defined in Figure 4, and v_{AB} and v_{BA} are found by setting θ equal to $\frac{\pi}{2} + \phi$ and $\frac{\pi}{2} - \phi$, respectively. Consequently, the times Δ_{AB} and Δ_{BA} become equal.

Primer that $\epsilon = \frac{1}{2}$ always owing to the principle of relativity (*SP*, pp. 88–89). This may well have been the part of the "analysis" that Bridgman mentioned in the passage from the notes of 1 August 1959, p. 10, that I quoted from previously, that proves that $\epsilon = \frac{1}{2}$. For the moment I postpone commentary on Bridgman's published proof which is incorrect because it is based on a space and time transformation whose spatial part displays the reciprocity of velocities between two inertial reference systems—that is, Bridgman did not include the conventionality of relative velocites.[14]

From January through early March of 1961 Bridgman wrote the first drafts of *Sophisticate's Primer*, in which he used the experiment from the notes of 1959 to argue at some length that a one-way velocity of light was not operationally definable. First of all, Bridgman noted the formula for the anisotropy of light, given in footnote 13, in which the measured transit times from A to B and from B to A were equal. More basically he realized that he had overlooked the point that the velocities of light from B to O and from A to O are two-clock velocities, as, he emphasized, are all velocities, because by definition velocity is the distance traveled in a time interval measured on two clocks that are situated at the beginning and end of the traversed spatial interval. Consequently a vicious logical circle ensues. As a result Bridgman wrote in the February 1961 manuscript of *Sophisticate's Primer* that Einstein had concluded correctly that "the question, is light *really* propagated isotropically?" can be answered only by posing an

14. In an epilogue to the first edition of *Sophisticate's Primer*, Grünbaum noted that Bridgman's conclusion was incorrect, but did not note where the error lay.

axiom system, or by definition if you prefer, because the isotropy of space for light propagation is linked to how clocks are synchronized.

Then on 7 April 1961 Bridgman inserted a page not intended for publication into the revised manuscript of *Sophisticate's Primer*. The inserted page is an interesting historical document, and in fact Bridgman very likely addressed it to posterity, since it instructs us how to read some of his unpublished notes of 1959, as well as the earlier versions of *Sophisticate's Primer*. We recall also that Bridgman was aware of his progressive illness. Bridgman wrote:

> In reading over any old material it is to be kept in mind that my thinking has experienced [certain] major changes. [One of them occurred] early in 1961, that "one-way" velocity of light can have no physical significance by itself, but is essentially a two-clock concept and has meaning only when the method has been specified by which time is spread over space. This realization negatived my efforts to find some physical method of measuring one-way velocity . . . .

This intellectual upheaval rendered the problem of measuring one-way light velocity "illegitimate" (*SP*, p. 47), because it involves a vicious logical circle. Thus, in the published version of *Sophisticate's Primer* after a brief description of the experiment in Figure 4 (*SP*, p. 47), Bridgman dismissed this measurement with short shrift (see also, *SP*, p. 48.)

What about the published incorrect proof in *Sophisticate's Primer* that $\epsilon = ½$? Could this proof have been the result of an error on Bridgman's part? Having studied Bridgman's extant manuscripts I would say yes, but not merely an error. Rather I should like to suggest that it was an act of desperation. For as we have seen from his notes, after much travail he had convinced himself of the experimental impossibility of measuring the one-way light velocity, and also of the necessity to assume that the one-way light velocity would always be c in order to do any research, for example, in astronomy; furthermore, he was convinced of the validity of the special relativity theory within a restricted domain of applicability. Thus he next had to divest a basic quantity of the theory—the velocity of light—of any conventional characteristics. In fact, if the one-way light velocity were a convention, then, as Bridgman wrote in the MS of 4 September 1960, the "principle of relativity itself must be recognized to be a 'convention.' " According to Bridgman, however, a convention is not a basic statement of a physical theory but is a "more or less incidental consequence of more deep-seated properties" (see Bridgman, 1927, p. 117). Here he was in a quandary about what to do, because he had to present a proof for the necessity and sufficiency of $\epsilon = 1/2$, so that "Einstein had no choice" (SP, p. 90).[15] Considering himself to be constrained to toe the operational line of sticking to the facts and not speculating in print on a quantity as basic as velocity, he offered the proof in *Sophisticate's Primer*, which is the one from the MS of 4 September 1960. Recalling that Bridgman usually refined and developed his writings further, I feel sure that he would have revised this proof in a later edition of *Sophisticate's Primer*. And almost certainly

15. In the revised MS of *Sophisticate's Primer*, Bridgman crossed out the statement that Reichenbach's "ϵ would have to be a function of v" in order to have ϵ be other than ½. This, in fact, is the case (see Winnie, 1970).

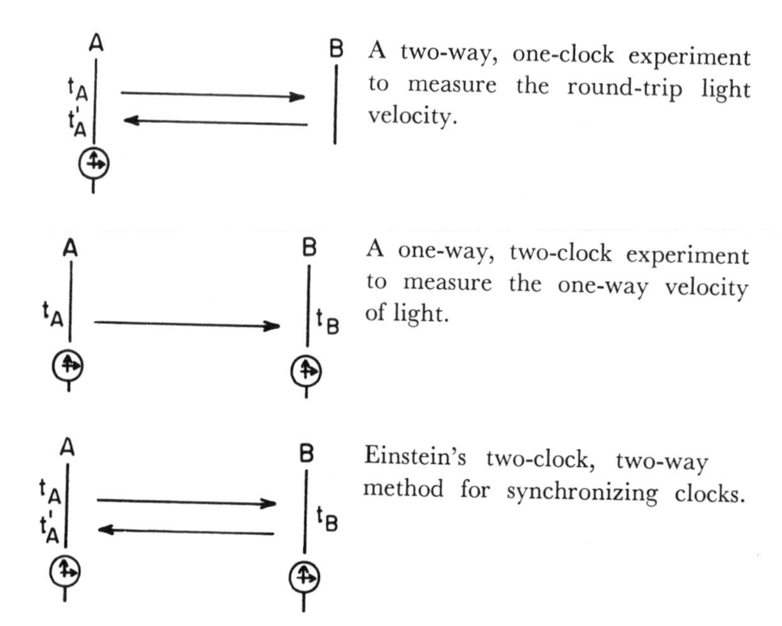

A two-way, one-clock experiment to measure the round-trip light velocity.

A one-way, two-clock experiment to measure the one-way velocity of light.

Einstein's two-clock, two-way method for synchronizing clocks.

he would have maintained that, whereas it is the case that there are alternate theoretical descriptions of phenomena, the individual scientist judges one among these theories to be of the greatest value because, as Einstein wrote in the 1905 relativity paper, it has removed "asymmetries which do not appear to be inherent in the phenomena."

From basic statements that are presumed to be independent of one-way velocity assumptions, John A. Winnie deduced ϵ-dependent Lorentz transformations that revert to the usual Lorentz transformations when $\epsilon = \frac{1}{2}$. (Winnie, 1970). Winnie rediscovered some of Bridgman's results on the conventionality of relative velocity and of length contraction—in particular, the conventional aspects of special relativity are rooted in real differences of clock synchronizations. Bridgman may well have glimpsed the sheer artificiality of the conventionality of distant simultaneity, which destroys symmetries of the sort that philosopher-scientists such as Einstein and Poincaré had in their own ways believed to be intrinsic to physical theories (see also Holton, 1973).[16]

The Observer

The first problem that Bridgman attacked in the manuscript of 11 July 1959 on the usual formulation of special relativity theory concerned the observer. As Bridgman put it, according to the operational view, "I am to be the central clearing house" (MS 11 July 1959, p. 1). This, in fact, is the way science is done when a single observer correlates meter readings whether he is performing a table-top experiment or correlating the registrations on instruments in different inertial reference systems through the application of the Lorentz transformations. Bridgman described this mode of formulating special relativity theory with the Machian term "minimum point of view" (*SP*, p. 163). Einstein, on the other hand, had assumed a "psychological attitude" that entailed several independent observers. In order to support his view that relativity can be addressed only by a single observer who correlates meter readings, Bridgman in *Sophisticate's Primer* cited the Terrell effect (*SP*, p. 130).

---

16. For example, for a certain value of ϵ the length contraction of a rod oriented along K''s x-axis and observed in K can be eliminated for the relative velocity of K' relative to K in the positive x-direction, but not in the negative x-direction; this is the case also for time dilation. In these cases ϵ is not the same in K and K'. Winnie also deduced the conventionality of infinitely slow clock transport.

In almost every exposition of special relativity theory, wrote Bridgman, the device of one or more observers is used "uncritically" to discuss how a moving rod appears. In 1959 James Terrell, by taking into account that light does not reach the eye simultaneously from different parts of a moving object, used special relativity theory to calculate that an observer does not "see" a Lorentz contraction but rather sees other sorts of distortions. Consequently, Bridgman concluded that since expositions and applications of special relativity theory had been carried out for over fifty years without awareness of the Terrell effect, then the device of observers traveling with frames of reference and discussing what they see is superfluous, since it turns out

(*opposite page*) A page from Bridgman's MS "Inertial Systems" where he proposed a measurement for the one-way light velocity. There are clocks and mirrors at A and B and a single observer at 0 who is also equipped with a clock. A light beam goes from A to B. An observer at 0 registers the time difference between the flash of light at A and the flash of light at B that denotes the beam's arrival at B to be

$$\Delta_{AB} = \frac{AB}{v_{AB}} + \frac{BO}{v_{BO}} - \frac{AO}{v_{AO}},$$

where $AB = 1$, $AO = BO$, and v_{AB}, v_{BO}, and v_{AO}, are the velocities of light in the directions AB, BO, AO, respectively. Similarly, for a flash of light emitted from B

$$\Delta_{BA} = \frac{BA}{v_{BA}} + \frac{AO}{v_{AO}} - \frac{BO}{v_{BO}},$$

where v_{BA} is the velocity of light in the direction BA. Bridgman defined the angle θ to be the angle between the direction of light propagation and the direction from A to B.

This suggets that we should go back and reexamine our assumption that operations are possible which will distinguish between the "go" and the "come" velocity of light

30/7/59

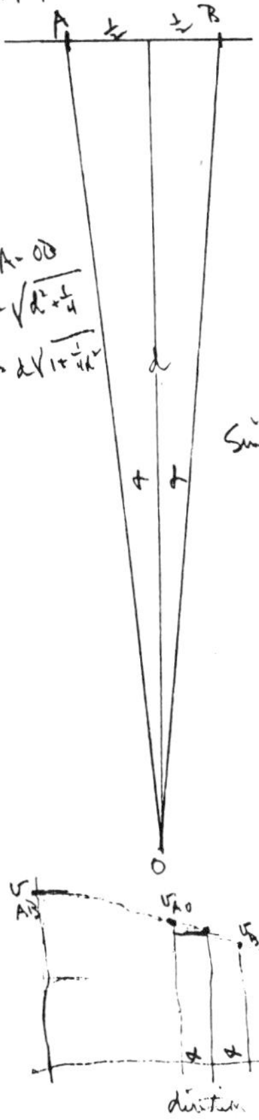

$OA = OB = \sqrt{d^2 + \frac{1}{4}} = d\sqrt{1 + \frac{1}{4d^2}}$

Given two points A & B unit distance apart a distant observing station O with clocks, & mirrors at A & B directed toward O to announce the arrival of light at A or B

Given first a beam of light from A to B. The observer at O times the difference between his observation of the arrival of light at A & B. Assume the velocity of light depends on the direction. Then the difference of arrival time of the beams is

$$\Delta_{AB} = \frac{1}{v_{AB}} + \frac{d\sqrt{1+\frac{1}{4d^2}}}{v_{BO}} - \frac{d\sqrt{1+\frac{1}{4d^2}}}{v_{AO}}$$

Similarly

$$\Delta_{BA} = \frac{1}{v_{BA}} + \frac{d\sqrt{1+\frac{1}{4d^2}}}{v_{AO}} - \frac{d\sqrt{1+\frac{1}{4d^2}}}{v_{BO}}$$

If these two Δ's are the same then the difference between the go & come times is not measurable and their equality is a convention

If v is a function of direction, then if we assume that they are in the order indicated, there is at least partial compensation, and if v is the proper function of direction it is conceivable that the compensation may be complete and $\Delta_{AB} \equiv \Delta_{BA}$ For

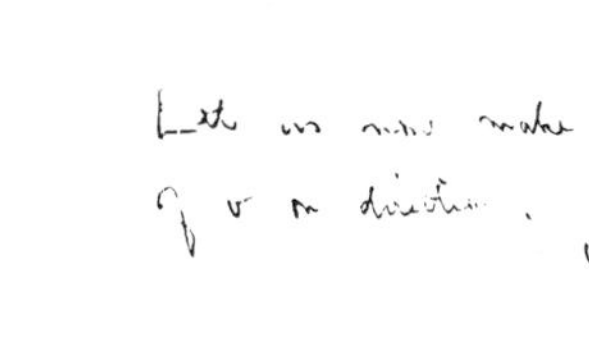

$$\Delta_{AB} = \frac{1}{v_{AB}} + d\sqrt{1+\frac{1}{4d^2}}\left(\frac{1}{v_{BO}} - \frac{1}{v_{AO}}\right)$$

$$\Delta_{BA} = \frac{1}{v_{BA}} - d\sqrt{1+\frac{1}{4d^2}}\left(\frac{1}{v_{BO}} - \frac{1}{v_{AO}}\right)$$

Let us now make a plausible assumption about the dependence of v on direction. Try

$$v_\theta = \frac{v_{AB}+v_{BA}}{2} + \frac{v_{AB}-v_{BA}}{2}\cos\theta$$

that an observer "may be deceiving" himself (*SP*, p. 130). Special relativity theory can be applied by a single observer at rest in the laboratory who correlates instrument readings from instruments either at rest or moving relative to him (*SP*, p. 151). For Bridgman these meter readings were the sole physical reality, and seeking these readings under every conceivable circumstance was "part of the ultimate task of the physicist" (*SP*, p. 152). Thus the electromagnetic field, and the propagation of light as well, are ensembles of meter readings, and a "field-in-the-absence-of-an-*instrument* has no meaning" (italics in original, *SP*, p. 157). This view of physical reality accorded Bridgman an instrumentalist definition of the electromagnetic field equations—they permit us to refine our meter readings (*SP*, p. 159). In agreement with Gustav Kirchhoff's widely quoted 1876 statement, Bridgman concluded that a "simple description, it seems to me, is the inevitable end of any analysis of which we are capable" (ibid.).

Consistent with his philosophy of science, Bridgman deemed that the role of the individual observer—the "I"—was essential to the enterprise of doing science. He wrote (1949) that science is "public" only to the extent that two or more observers can agree on a phenomenon. But, he emphasized, "it is a matter of simple observation that the private comes before the public." In my opinion, Bridgman's strong belief and life-long struggle to maintain the "I" in the scientific enterprise lay behind his never accepting general relativity theory. Most likely as a result of not having read Bridgman's earlier writings, Einstein (1949) missed this point when he dismissed Bridgman's at times strongly put and not always substantively based criticisms of general relativity theory with the comment that "to consider a logical system as physical theory it is not necessary to demand that all of its assertions can be independently interpreted and 'tested,' 'operationally.' " Here Einstein addressed himself only to the part of Bridgman's critique that concerned Einstein's transference of physical reality from space-time intervals to the metric tensor. Bridgman's deep scientific-philosophic concerns about general relativity were: (1) that it focused on the coordinate system rather than on events themselves because there were no standard measuring clocks and rods as in special relativity theory; and (2) that "certainly the feeling that one gets from reading many of the fundamental expositions, even those of Einstein himself, is that this business of getting away from a special frame of reference and observer to something not fettered to a special point of view is very important," i.e., the general relativity theory's emphasis on covariance (1936). This led Bridgman (1949) to conclude: "Perhaps the most sweeping characterization of Einstein's attitude of mind [i.e., realist philosophy] with regard to the general theory is that he believes it possible to get away from the special point of view of the individual observer and sublimate it into something universal, 'public,' and 'real.' " Bridgman's opinion of realism versus operationalism in the 1940s was very likely the one that appears in *Sophisticate's Primer* (pp. 154–155) and that lies at the crux of his concern regarding the general theory—i.e., that it excluded the observer. He had expressed this point most strongly in 1936:

> It cannot be too strongly emphasized that there is no getting away from preferred operations and a unique standpoint in physics; the unique physical operations in terms of which interval has its meaning afford one example, and

> there are many others also. There is no escaping the fact that it is *I* who have the experiences that I am trying to coordinate into a physical theory, and that *I* must be the ultimate center of any account which I can give. I and my doings must be specially set apart and, in perhaps the only possible sense of the word, constitute an absolute. It seems to me that to attempt to minimize this fact constitutes an almost wilful refusal to accept the obvious structure of experience. In so far as the general spirit of relativity theory postulates an underlying "reality" from which this aspect of experience is cancelled out, it seems to me to be palpably false, and furthermore devoid of operational meaning.

Furthermore, in the light of this quotation we can understand Bridgman's additions to his notes of 1 August 1959—concerning all "our [my]" analyses (see p. xxx of this Introduction).

Bridgman's emphasis on the "I" was sometimes criticized as solipsistic. His calling attention to the Terrell effect in *Sophisticate's Primer* can be taken to be an explicit science-based defense of the consistency of maintaining the point of view of a single observer who reads instruments. With the forthrightness that had become his trademark, in the *The Way Things Are* (1959b) he defended the operational philosophy against the charge of solipsism. (Unfortunately the Terrell effect was as yet unknown when he wrote that book.) Here Bridgman wrote that "there *has* to be a plurality of sorts because there are many people, and there *has* to be a dichotomy of sorts" because there is, for example, your-science and my-science. This necessary plurality and dichotomy he took as grounds for not uncritically accepting any unitary world view, even the positivistically inclined Unity of Science movement that emphasized public science. The problem that Bridgman faced was how to fit this plurality and dichotomy into "my picture of the world," without being accused of solipsism. In *The Way Things Are* he provided a Gordian solution of his own. Bridgman dismissed as "innocuous" the unavoidable pluralistic aspects of "viewing everything from myself as center," because this description of nature gives us the "only unity we can use, the only unity we need, and the only unity possible in the light of the way things are."

Although one may not always agree with the operational point of view, I feel sure that students and teachers alike will pay Bridgman's final book on the philosophy of science that highest compliment of picking up a pencil in order to follow better his train of thought, and to write their own views in the margins as a result of their having been inspired to "stand back a little for critical scrutiny of what he has really got."

Bibliography

Bridgman, Percy W.

Manuscripts:

27 December 1922	Preliminary for Space Time Symposium.
September 1923	Critical Discussion of Relativity.
11–18 July 1959	Reformulation of Special Relativity.
24 July – 2 August 1959	Inertial Systems.
18 August 1959	Comment on Ives' Papers on Relativity.
26 May 1960	Distance as a 'Convention.'
4 September 1960	Two Footnotes to Special Relativity Theory
January–April 1961	Manuscripts of *Sophisticate's Primer*

Published Materials:

1924 A Suggestion as to the Approximate Character of the Principle of Relativity, *Science*, *59*, 16–17 (1924).

1927 *The Logic of Modern Physics* (New York: Macmillan, 1927).

1936 *The Nature of Physical Theory* (New York: Dover, 1936). Based on three Vanuxem lectures delivered during December 1935 at Princeton University.

1938 Operational Analysis, *Philosophy of Science*, *5*, 114

(1938); reprinted in P. W. Bridgman, *Reflections of a Physicist* (New York: Philosophical Library, 1955), pp. 1–26.

1949 Einstein's Theories and the Operational Point of View, in P. A. Schilpp (ed.), *Albert Einstein: Philosopher-Scientist* (Evanston, Ill.: Library of Living Philosophers, 1949), pp. 333–354.

1954 Remarks on the Present State of Operationalism, *Scientific Monthly*, 79, 224–226 (1954); reprinted in *Reflections of a Physicist*, pp. 160–166.

1959a The Logic of Modern Physics after Thirty Years, *Daedalus*, *88*, 518–526 (1959).

1959b *The Way Things Are* (Cambridge, Mass.: Harvard University Press, 1959)

1962 *A Sophisticate's Primer of Relativity* (1st ed. Middletown, Conn.: Wesleyan University Press, 1962).

Einstein, Albert

1905 Zur Elektrodynamik bewegter Körper, *Annalen der Physik*, *17*, 891–921 (1905); translated in Miller (1981).

1911 Zum Ehrenfestschen Paradoxen, *Physikalische Zeitschrift*, *12*, 509–510 (1911).

1923 Fundamental Ideas and Problems of the Theory of Relativity. Lecture delivered on 11 July 1923 to the Nordic Assembly of Naturalists at Gothenburg, in acknowledgement of the Nobel Prize. Reprinted in *Nobel Lectures, Physics: 1901–1921* (New York: Elsevier, 1967), pp. 479–490.

1949 Reply to Criticisms, in P. A. Schilpp (ed.), *Albert Einstein: Philosopher-Scientist* (Evanston, Ill.: Library of Living Philosophers, 1949), pp. 665–688.

Grünbaum, Adolf

1962 Prologue and Epilogue to P. W. Bridgman, *A Sophisticate's Primer of Relativity* (1st ed., Middletown, Conn.: Wesleyan University Press, 1962).

1973 *Philosophical Problems of Space and Time. Boston Studies in the Philosophy of Science, 12* (Dordrecht: Reidel, 1973).

Holton, Gerald

1973 *Thematic Origins of Scientific Thought: Kepler to Einstein* (Cambridge, Mass.: Harvard University Press, 1973).

Ives, Herbert E.

1951 Revisions of the Lorentz Transformations, *Proceedings of the American Philosophical Society, 95*, 125–131 (1951).

Kemble, Edward (with Francis Birch and Gerald Holton)

1970 Bridgman, Percy Williams, in C. C. Gillispie (ed.), *Dictionary for Scientific Biography, 2*, 457–461 (1970).

Mach, Ernst

1960 *The Science of Mechanics: A Critical and Historical Account of its Development* (1st German edition 1883; LaSalle, Ill.: Open Court, 1960), translated by T. J. McCormack.

Miller, Arthur I.

1978 Visualization Lost and Regained: The Genesis of the Quantum Theory in the Period 1913–1927, in J. Wechsler (ed.), *On Aesthetics in Science* (Cambridge, Mass.: MIT Press, 1978), pp. 72–102.

1981 *Albert Einstein's Special Theory of Relativity: Emergence (1905) and Early Interpretation (1905–1911)* (Reading, Mass.: Addison-Wesley, 1981).

1982 Redefining *Anschaulichkeit*, in *Festschrift for Laszlo Tisza* (Cambridge, Mass.: MIT Press, in press 1982).

in press Poincaré and Einstein: A Comparative Study, in press A. I. Miller, *On the Nature of Scientific Discovery* (Cambridge, Mass.: Birkhäuser, in press).

Poincaré, Henri

1898 La mesure du temps, *Revue de Metaphysique et de Morale, 6*, 371–384 (1898); reprinted in H. Poincaré, *Value of Science* (Paris: Flammarion, 1905; New York: Dover, 1958), pp. 26–36, translated by G. B. Halsted.

1902 *Science and Hypothesis* (Paris: Flammarion, 1902; New York: Dover, 1952), translator unknown.

Reichenbach, Hans

1958 *The Philosophy of Space and Time* (Berlin: Gruyter, 1924; New York: Dover, 1958), translated by M. Reichenbach and J. Freund.

Winnie, John A.

1970 Special Relativity without One-Way Velocity Assumptions, *Philosophy of Science*, I, *37*, 81–99; II, 223–238 (1970).

1972 The Twin-Rod Thought Experiment, *American Journal of Physics*, *40*, 1091–1094 (1972).

A Sophisticate's Primer of Relativity

Introduction

In writing this monograph I have in mind the reader who already has a first acquaintance with the subject matter of special relativity, and who feels the need to stand back a little for a critical scrutiny of what he has really got. Many students must have felt such a need, for in the conventional expositions of the subject many topics are left without the critical examination which is plainly called for. The need for a critical scrutiny of fundamentals is underlined by the failure of even seasoned relativists to reach consensus, as shown by the continuing series of controversies on fundamental matters; the latest of these, with regard to the space traveler, has hardly yet subsided.

Among the questions which naturally present themselves and are subject to detailed examination here are: What is a meter stick? What is a clock? What is a frame of reference? What is the role of the "observer" — is he to be identified with some frame of reference or is he outside of any frame? What is an "event"? Can the "conventional" element in the "definition" of distant simultaneity be avoided? What is a "law of nature"?

Every student of relativity, even one who has only a

first acquaintance with the subject, doubtless has answers of sorts to these questions, which he gives on the basis of how he sees relativity theory handling itself in specific situations. But answers the student obtains from what he reads into the theory are likely to be distorted, in that more is read into the theory than is necessary. It seems to me important that our answers to these questions should be minimum answers, in the sense that no more should be read into the theory than is necessary for the coherence and stability of the whole structure. In the following, special effort is made to conduct the discussion from such a minimum point of view. This minimum point of view I believe is particularly demanded in discussing the relation of the observer to the frame of reference. Practically all elementary expositions are given in terms of different observers, each one associated with his own frame of reference. This I believe is going further than is necessary and involves implications inconsistent with what the theory is really saying, implications we shall try to avoid.

In the exposition, an early chapter is devoted to phenomena in a single frame of reference. We then consider how these phenomena appear in two frames, one moving with respect to the other, which provide the conventional subject matter. The reason for this approach is that there are some issues which present themselves in the context of a single frame — for example, distant simultaneity and the setting of distant clocks — which nevertheless constitute part of the essential subject matter of relativity. Such issues were always present before the advent of relativity theory, and the critical insight of Einstein should not have been required to make us aware of them.

The "critical" attitude of this essay is critical only to

the extent that it is necessary to find out what the theory essentially is. The theory as ordinarily understood is accepted without question as a working tool which, in its ostensible universe of discourse, gives the best control we have yet been able to acquire of phenomena. This universe of discourse reaches down into the microscopic domain of subatomic phenomena. In fact, this control is so good that no phenomena now in sight would seem inconsistent with the theory within present limits of experimental error. But even if we accept the theory in this spirit, it does not mean that all problems of understanding have been solved. This essay is directed toward these problems of understanding.

The Lorentz Transformation Equations

A number of writers like to define the content of the special theory of relativity as coextensive with the content of the Lorentz equations. For reasons that will appear as we progress, I believe this to be only a partial characterization of the situation. Nevertheless, many aspects of the situation are covered by the equations, and it will be convenient to begin with them. Two aspects of the equations are to be distinguished: an abstract mathematical aspect and a physical aspect which appears when the equations are applied to physical situations.

Mathematical Aspects

It will be sufficient for our present purposes to consider the equations in their simplest form, in which we are concerned with only a single spatial dimension. The equations are:

$$\left.\begin{aligned} x' &= \frac{x - vt}{\sqrt{\left(1 - \frac{v^2}{c^2}\right)}} \\ t' &= \frac{t - \frac{vx}{c^2}}{\sqrt{\left(1 - \frac{v^2}{c^2}\right)}} \end{aligned}\right\} \tag{1}$$

The full three spatial dimensions can be covered if desired by simple addition of the two equations:

$$y' = y;\ z' = z.$$

In their mathematical nakedness, these equations are formulas by which a pair of numbers, x' and t', may be calculated, given any four other numbers x, t, v, and c. Since we shall be interested only in real numbers, we accordingly restrict the arbitrary values permissible to give to x, t, v, and c. This means, in particular, that we restrict v to be less than c.

The equations may be inverted, solving for x and t in terms of the four other quantities:

$$\left.\begin{aligned} x &= \frac{x' + vt'}{\sqrt{\left(1 - \frac{v^2}{c^2}\right)}} \\ t &= \frac{t' + \frac{vx'}{c^2}}{\sqrt{\left(1 - \frac{v^2}{c^2}\right)}} \end{aligned}\right\} \tag{2}$$

Inspection shows that the inverted equations can be obtained from the original equations by changing the sign of v and switching the position of the primes.

In general, these equations impose two conditions on six quantities, such that when any four of the six are assigned arbitrary values within the imposed limits, the other two are thereby fixed.

The equations have certain purely mathematical properties. Among these properties are the following:

(1) Whenever both x and t are zero, both x' and t' are zero, whatever v and c.

(2) When $x' = 0, \; x = vt$.

(3) When $x = 0, \; x' = -vt'$.

(4) Whenever $dx/dt = c, \; dx'/dt' = c$, whatever v.

(5) When $t = 0$ and $x' = 1, \; x = \sqrt{\left(1 - \frac{v^2}{c^2}\right)}$.

(6) When $x' = 0$ and $t' = 1, \; t = \frac{1}{\sqrt{\left(1 - \frac{v^2}{c^2}\right)}}$.

(7) When $x'/t' = u, \; x/t = \frac{u + v}{1 + \frac{uv}{c^2}}$.

The essence of the equations is in a certain sense contained in these properties, for it can be shown that, given the properties, the equations are determined. In fact, not all the properties are needed to determine the equations, so that from this point of view some of the properties are redundant. In particular, a simple algebraic analysis shows that the equations as given are the only linear relations connecting x' and t' with x and t which satisfy the first four conditions and also the symmetry condition that the inverted equations can be obtained by changing the sign of v and switching the position of the primes. From this point of view, therefore, conditions (5), (6), and (7) are redundant. It is an interesting question, which should be answerable by straightforward analysis, to find to what extent any of the four conditions selected above can be replaced by any of the other three conditions.

All seven properties of the equations obviously have physical significance. For example, it can be shown in a few lines that condition (4), which we shall see later has the physical significance that the velocity of light is invariant, can be replaced by condition (7), which physically is the addition law for velocities. Condition (4), which is in differential form, can be replaced by the condition in finite form: "When $x/t = c$, $x'/t' = c$, whatever v." This condition and the corresponding condition on the derivatives are physically equivalent under the special conditions to which the postulates of relativity apply.

The form of the equations naturally suggests to the mathematician that he treat the four quantities x, t, x', and t' as "coordinates," the first pair in one coordinate system and the second in another. The remaining two quantities v and c are then to be treated as "parameters." It is not mathematically necessary to ascribe these connotations to these quantities, but if the connotations are accepted, the equations lend themselves to a simple geometrical interpretation. In fact, as was first shown by Minkowski, the equations become simply the equations for transforming from an orthogonal set of axes, x and t, to an oblique set, x' and t', at the same time introducing a suitable change of scale along the t and t' axes. The oblique axes are inclined to the original orthogonal axes at an angle $\tan^{-1} \frac{v}{c}$. By utilizing this geometrical interpretation it is possible to give great vividness to the content of the equations, and most writers on relativity make extensive use of the geometrical interpretation. In fact, it is usually considered that Minkowski's discovery of his transformation was one of the most important contributions to relativity theory after Einstein's original contribution. However, there is danger that, by emphasizing the geometrical properties of the transformation, an attitude will be engendered in which the physical significance of the equations will be obscured. This attitude is made easy by Minkowski's often quoted remark that henceforth space and time become shadows of the only reality, the four-dimensional manifold of space-time. Not all writers, however, regard the vividness attained by the four-dimensional representation as an unmixed blessing. For example, H. Dingle* says: "A disaster of the first magnitude (from the point of view of the understanding, as distinct from the extension, of the theory) occurred in 1908, when Minkowski transferred the subject from the realm of physics into that of mathematics."

Physical Aspects

So much for the naked mathematics of the Lorentz equations. We turn now to the physical applications. Nothing explicit in the equations themselves determines the nature of the physical application, but this has to be specified in some way apart from the equations. Not until we have specified the details of the physical application do we have the right to speak of the equations as part of a physical "theory." The machinery by which the physical significance of the

* H. Dingle, *Phil. of Sci.*, **27**: 253 (1960).

symbols in the equations is fixed has been described in various terms. I have spoken of the "text" which always accompanies a system of equations; Dingle speaks of "rules of correspondence," Margenau of "epistemic correlations."

Whatever way we talk about it, we follow the mathematician in taking x, t, x', and t' as "coordinates," x and t in one coordinate system and x' and t' in another. We call these systems the "stationary" and the "moving" respectively. Now physically we never have simply coordinates; we always have coordinates *of* something. There is a technical name in relativity theory for this thing that the coordinate is *of*; it is called an "event." Physically, the nature of this event is by no means determined, but a great deal of latitude is permissible. For example, the arrival of an electron or a proton or a photon at a specified place at a definite time are all to be indifferently described as "events." Whatever the physical nature of the event, the equations are concerned with only one aspect of it and in fact are *capable* of handling only one aspect of it. This aspect is the coordinate aspect, namely the x-coordinate of the place where the event occurred and the t-coordinate (that is, the time) at which it occurred. (Any complications arising from the three coordinates of space need not detain us.) Given now the coordinates, the event is by no means determined. We might be dealing with either electron or proton or photon. Which of these it is, or whatever else it might happen to be, has to be specified in the "text."

Given the event, we assume that there is some procedure by which its coordinates (x, t couple) may be determined. But before we can have the coordinates we must have the event, and this must be specified and recognizable in some terms other than through the coordinates. That is, the system of events, with which our equations or our theory are concerned, is something prior to the coordinates, existing in its own right and independently describable. We partially describe this system of events by giving the numerical values of the coordinates in one or another coordinate system. From the point of view of the coordinate system (which we will speak of alternatively as a "frame of reference"), there is, therefore, a touch of the "absolute" about an event. This remark is justified by the simple observation that we speak of the "same" event occurring in both stationary and moving systems.

Assuming now that we have our self-contained system of events, we must inquire in detail by what method we assign coordinates to them. This method involves some sort of physical procedure; eventually it must be such that it will give us coordinates in both the stationary and the moving frames of reference. But before we have two coordinate systems we must have one, and issues arise in connection with a single frame of reference which must be solved before we can pass to two. In the next chapter, we consider issues peculiar to a single frame of reference.

Single Frames of Reference

Frames of Reference in General

It would be well to begin by considering what the function of a frame of reference is and under what circumstances even a single frame is necessary. Nature does not present us with frames of reference ready made and labeled as such. We construct them for our own purposes out of the objects we find. To the naïve, primitive intuition, which is satisfied with mere recognition of the objects of its experience as they occur and recur, a frame of reference is not necessary. The necessity for a frame does not arise until one attempts to analyze one's experience to the extent of at least describing it. Certain properties of the objects of our experience can be satisfactorily described without the use of a frame of reference, but others demand a frame. Everyone understands what is meant when it is said that a certain object is "red," but no acceptable information is conveyed by the unqualified statement that the object is "there." Certain attributes of the objects of experience, such as position, velocity, or direction of motion, can be adequately described only in the context of a frame of reference. The attributes which thus demand a frame of reference may be among the simplest attributes of the objects of experience. Our frame of reference may be more or less complex, depending on the system and the attribute with which we are dealing.

Perhaps the simplest of all frames presents itself when we deal with a single pair of bodies subject to no influences except their own interaction. Here the frame of reference is simply the line connecting the two bodies. In this frame, we can describe the distance apart of the two bodies (that is, their relative position) and their relative velocity (that is, the time rate at which their distance of separation changes). In this simple frame of reference, there is a self-contained mechanics — there is a single mass parameter for the two bodies together which determines how the distance of separation accelerates when a force acts between the members of the pair, and there is a conservation law of energy in terms of the two parameters.

For more complicated systems of objects, this simple frame of reference is not adequate. We require something which will permit us to single out any particle and describe its motion in vector terms. For this, the ordinary coordinate system of the mathematician is adequate if we amplify it by adding some means for determining the times at which the various events in which we are interested occur. This implies a clock of some sort. Such a coordinate system itself has to be anchored in some way so that we can return to it when necessary. For this purpose, the coordinate system usually is associated with our earth for terrestrial experiences or with the solar system for astronomical experiences. Such a coordinate system is what we shall understand in the following by a "frame of reference."

Of course, it would be possible to split metaphysical hairs here and say that when we talk about an object being "red" we are tacitly using a frame of reference, because such a statement is meaningless without the "frame of reference" of all our past experience. However, we refuse to entangle ourselves in such verbal matters and shall understand by "frame of reference" what is usually called a coordinate system, with time added.

Significance of the Coordinates in a Single Frame

In the discussion in this chapter of a single frame of reference, we shall at first simplify by using a coordinate system in which there are only two coordinates, a single geometrical coordinate and the time coordinate. In this single coordinate system, x and t denote the place and the time at which some specified event occurs. We specify the place by giving a number which marks the location on the x-axis where the event occurred. We specify the time by giving the numbers on the dial of a clock to which the hands pointed when the event occurred. The numbers of the x-axis are previously assigned so as to measure the distance from an arbitrary point on the axis we choose as origin. The x-axis is thus itself a physical system of some sort, such that it has identifiable points which mark the position of the origin and the other coordinate numbers. We may think of it as a "rigid" rod, on which the distances of its various points from the origin are marked once for all, so that when an event occurs at some point along the rod we can locate it unambiguously by giving the number of the point on the rod at which it occurs. We do not usually scrutinize with any particular care the details of the process by which we set up our axis and number its points. Ordinary experience indicates that the details do not make much difference, provided only that we pay due regard to the considerations that common sense tells us are important. Thus we take care not to make our meter sticks of rubber or other easily deformable material, or to make our calibration on an excessively hot or cold day, when we know that thermal expansion in the meter stick is something to be reckoned with. "Rigidity" is a concept at which we arrive by a method of successive approximations we seldom take the trouble to analyze carefully, but which we have no reason to think, within the realm of ordinary experience, is not capable of complete logical self-consistency. That is, ordinarily we take for granted our meter sticks and our method of using them in measuring distance. For the present we leave it with this remark, but we shall return again to an examination of the question "What is a meter stick?"

Time of Distant Events

The time of the event is to be measured with a "clock." Again we see that we must sometime consider the general question "What is a clock?" For the immediate present we need consider only a single aspect of the general question, an aspect which does not present itself when measuring distance with meter sticks. The time of an event which happens here, at the origin, can be determined without question by simply observing the reading of the clock here when the event occurs. That is, for the present purposes we do not feel the need of analyzing the process by which we determine the time of events which take place here. But the event may occur at a distance. How shall we determine the time we are to associate with its x? We assume that we can see from here the x at which the event occurs, or an assistant may report it to us. But its time? We can observe on our clock here the time at which we see it occur there. This is what we ordinarily do, and within the world of ordinarily perceived phenomena it is perfectly satisfactory. But when we acquire sufficient skill to make our measurements much more sensitively than our unaided sense organs could possibly permit, we discover that there are propagation effects connected with seeing things happen, and we realize that we should correct our reading of the time at which we see the distant event happen for the time required by light to travel from there to here. But how to make this correction? It will evidently involve the velocity of light, but this is not easy to measure.

We can think of another method of determining the time of a distant event. We might have an assistant stationed where the event is expected to take place, and instruct him to observe and report to us the time he reads on a clock situated where the event occurs. We may suppose this local clock to be a watch our assistant has carried with him, taking pains of course that this watch be a timepiece of the utmost reliability. We do not, in fact, actually need an animated assistant, for the whole operation, including transmission of the local time of the event to us, could be done with automatic instruments. This suggested procedure does not involve the velocity of light in any obvious way and is therefore one we might be at first inclined to adopt. Furthermore, we believe from present evidence that if we refine the sensitiveness of our measurements only to such a point that the corrections we make for the velocity of light are accurate only to terms of the "first order," we shall find that the two methods (that is, correcting the local clock for light velocity and transporting a clock to a distance) agree.

But, going further, the present indications are that when we increase the sensitiveness of our measurements so that we can make corrections to what we shall later call "second-order" terms, we shall find that the two methods do not agree. In fact, we shall find that the second method described above is not internally consistent, for different assistants of ours, who have gone to the distant station by different routes, do not agree in the time which they report to us for the distant event. We find this disagreement between our assistants so disturbing that we are inclined to abandon altogether the second method for timing a distant event, and use only the first method, correcting as well as we can for the velocity of light. This is, as a matter of fact, the course almost universally adopted. Ives is the only writer known to me who has seriously tried to retain the second method. We shall return to this topic later.

Rates and Settings of Clocks

To see a little more clearly what is involved when we talk about the time of events at different places, we shall have to examine more carefully the notion of "clock." Our ordinary timepieces have two variable parameters; the "rate" may be controlled, and the hands may be "set" so as to start from any arbitrary zero. When we set up clocks at different places, we are concerned with both the rates and the settings of the separated clocks. The rates of clocks at different places may be compared by direct experiment to find whether they are equal. Many different methods may be used for comparing rates. We may, for example, send a light signal from the clock here to the clock there and note the difference between the time of departure on our clock here and the time of arrival at the clock there. If this difference of times is always the same when we repeat the experiment, the "rates" of the two clocks are the same. Or we may compare the rates in terms of acoustical signals sent from one clock to the other. If the rates are equal by the optical method of test, they will also be equal by the acoustical method. Furthermore, it seems to be a matter of experiment that a clock transported from here to a distance, where it is again established at rest, continues to run at the original rate. Accordingly, in all our future discussion of "spreading time through space" (that is, timing distant events), we shall always assume that the clocks at different places are running at the same rate. This is something over which we exert no control. We can and shall, however, exert control over the settings of the clocks at different places; it is this that we have in mind when we talk about spreading time over space in different or arbitrary ways. There are as many ways of spreading time over space as there are ways of setting clocks in different places.

In setting separated clocks, some optical procedure is usually specified. This, however, is not necessary; just as in comparing the rates of clocks in different places either an optical or an acoustical procedure may be adopted, so in setting separated clocks either an optical or acoustical or other procedure may be adopted. If, for example, the event is one we can hear, such as the stroke of a distant woodsman's axe, then we may correct the time at which we hear the distant event occur for the time required for the propagation of sound. Or, as before, we could set a clock locally beside the woodsman in such a way that the time at which we *hear* the event occur on the distant clock (perhaps by the use of a suitable arrangement of chimes) agrees with the corrected time of our local clock. Of course, the correction I apply to my local clock for the propagation of light is not the same as the correction I apply for the propagation of sound (one can *sense* that the sounds of the distant axe strokes are not synchronous with the strokes one sees), but, nevertheless, the clock at the distant station which I set by what I see agrees with the other distant clock I set by what I hear.

The point made in the last paragraph is, I think, not usually appreciated. The usual impression seems to be that because light often plays an important role in setting distant clocks, light is essential. Light is actually used because it is widely applicable in ordinary situations; and moreover, in the total range of physical conditions we wish to cover, as when we later pass to moving frames of reference, the acoustical method becomes inconveniently complicated because we have to take account of the medium through which sound is transmitted. But, "in principle," acoustical signals or any other sorts of mechanical signal could be used to set distant clocks, and all methods, including the optical, would give identical results.

Spreading Time over Space

What we effectively did above, when we specified how to set the clock at a distant station by light or by sound, was to set up one or another special method of spreading time over space. Now if we are to talk meaningfully at all about the "time" of distant events, we must adopt *some* method of spreading time over space. (This holds even if we elect to time the distant event by applying a suitable correction to the reading of our local clock. Such correction is a function of the place at which the event occurred, and in finding the best way of correlating the correction with the location we have effectively spread time over space.) There are certain properties of *any method whatever* of spreading time over space that now demand our attention.

In dealing with time measurement, we at once encounter the notion of time intervals, or durations. By definition, a time interval, or duration, is the difference of two times — that is, the difference of two clock readings. Intervals, or durations, imply two events between which we are getting the time interval. If the two events occur here, the time interval between them is simply the difference of the time readings on our local clock. If the two events occur at different places, then the interval is, *by definition*, the difference of the two clock readings on the two clocks at the two places where the two events occurred, or the difference of two suitably corrected readings on our single local clock. Instead of two isolated events at a distance from each other, we may be dealing with a chain of connected events, with a recognizable initiating event and a recognizable terminating event. This chain of events may be all local, or it may be spread over a distance, as, for example, in a journey. Such a spread-out, compound event may be decomposed into a succession of component events. It is a property of such compound events, whether confined to one place or spread out, that the sum of the time intervals or durations of the component partial events is equal to the total duration of the compound event. The proof is obvious, for on summing the durations, each intermediate clock reading occurs twice, with opposite signs, once at the end of any particular episode and again as the beginning of the succeeding episode. All the intermediate times cancel in pairs, therefore, leaving only the initial and final times of the compound event. This additive property of the component durations of a compound event evidently holds *no matter how time is spread over the space of the intermediate events.* This observation will prove to be of the greatest importance when we consider the special properties of light. It follows, in general, that we can never infer from a measurement of the duration of a compound event what was the duration of any intermediate component event. In particular, if the compound event consists of a chain of events spread out through space but returning to the starting point, one can never infer from a measurement of the total duration on the single clock at the place of beginning and end what was the manner in which time was spread through space at the intermediate points, or what was the duration of any of the component episodes.

That the considerations of the last paragraph have wide applicability appears when we reflect that we use only intervals or durations in every description of the temporal aspects of phenomena, the "absolute" time being by hypothesis and experiment irrelevant. Consideration of other aspects of the measurement of time we leave until later.

Velocity

Next after position and time, we have to measure "velocities" in the stationary system. In measuring velocity, most of the issues recur that arose in determining the time of distant happenings.

In the everyday world of direct sense perception, the

phenomena of motion are universal and important, and part of man's task of adapting himself to his environment is to adapt himself to the phenomena of motion. The phenomena of motion in their entirety exhibit great complication, but, nevertheless, adequate mastery has been acquired by many. The aborigine controls his boomerang, the baseball pitcher his curve ball; and every small boy has found how to adjust the direction in which he throws his stone and the speed with which he throws it to have the greatest probability of attaining his objective — the intersection at the same instant of the trajectory of the stone with the trajectory of the fleeing cat.

Motion as such is evident to direct perception — we *see* that this or that object is moving, with no conscious analysis of the perception. But when we do attempt to analyze further, as perhaps when we try to *measure* the velocity of some object, we encounter difficulties which sometimes still make trouble. We have been less successful in dealing with motion and velocity conceptually than in dealing with them practically. A large part of the resistance to acceptance of the so-called paradoxes of relativity theory arose from an imperfect comprehension of what is involved in the simple concept of motion. Long before Einstein, the early Greek philosophers had found difficulty here and were genuinely puzzled by such things as the paradoxes of Zeno. It is not easy for us from this distance to appreciate properly the reason for their difficulty. Perhaps it was because they took too seriously the direct evidence of their senses, which lends itself to the idea that the world can be exhaustively described at any moment in terms of the *positions* of its component parts. It took the insight of Galileo to appreciate that the world is not describable in terms only of the two parameters position and time, but that a third parameter, namely velocity, is necessary. At any instant, a body *has* velocity as well as position, and it is necessary to give both position and velocity to describe it fully.

But how to *measure* this velocity? A moving body changes its position with time; if it does not, we call its velocity zero. It is therefore natural to try to measure its velocity by specifying how much its position changes in some definite time. The natural way to do this is to observe the time at which the body is here and the later time when it has moved a known distance away from here, and divide the change in position (distance moved) by the time consumed. But to do this we have to know at what time it was at a distance, and the same questions arise that arose when we were trying to specify the time of distant events. We may either observe here the time at which we see (or hear) the body arrive at its distant position — in which case we have to correct for the time of propagation of light (or sound) from there to here — or we or our assistant may observe the time of arrival on a clock at the distant station. In the latter case, we have to solve the problem of the best method of establishing a clock at the distant station.

If we have adopted a method for timing events at a distance, we can also specify a method for measuring velocity, namely to divide the distance traveled by the difference of times as recorded here when the object left and there when it arrived. In whatever way we measure the time of distant events, the concept of velocity at which we thus arrive is a "two-clock" concept. This is obviously the case if we use a clock here and a clock there to time the body over its course, but it is also the case if we determine the two times on a single clock here and make correction for the time required by light to travel from there to here. For in making this correction we have to measure the velocity of light, and this measurement demands a second clock. But if velocity is a "two-clock" concept, what becomes of Galileo's insight that velocity is something a body *has now?* In other words, is it possible to measure the velocity of a body *now*, at this instant, without somehow involving two different clocks? In the world of everyday perception this *is* possible, because there are instruments which give the velocity *now* directly on a dial. The speedometer on the dashboard of our car is such an instrument. But when we analyze down to time intervals short enough to take us out of the world of direct perception, this sort of answer becomes less satisfactory. For analysis shows that what any such instrument gives is a somewhat blurred average of the condition of the instrument through a time interval of the order of that required for elastic disturbances to travel through the members of the instrument.

In general, when we push our analysis far enough, we find that "here-now" becomes fuzzy. The difficulty we are encountering is not peculiar to the concept of velocity. Any activity of ours involves distributions in space and averages in time; in particular, this applies to the operation of the nervous machinery which gives us our perceptions.

Returning to Galileo's insight, when we consider phenomena in their entirety, including the phenomena of mechanics as well as those of kinetics, the third instantaneous parameter appears. Actual bodies have momentum, in virtue of which they may exert pressures, which may be directly measured in terms of elastic distortions at an instant. In the ranges open to direct perception, these pressures are manifest to the sense of touch. Perhaps Zeno's difficulty was connected with trying to give a complete account of the world in terms of a single sense, the sense of sight. If he had been willing to include touch on the same basis, possibly his difficulties might not have arisen.

So far, we have been considering only the "two-clock" aspect of the velocity of everyday experience. I apply this concept in describing the bodies I see moving about — automobiles or rifle bullets or sputniks. But *I* also, as well as external objects, move about, and in so doing experience "velocity." I may, then, speak of two kinds of velocity, a "my-velocity" and a "your-velocity." I can and do measure "my-velocity." I do this by combining the readings of a watch I carry in my pocket with the distances along my road. These distances I may have previously laid out, or if I have not been foresighted enough to measure the distances previously, I can measure them as I proceed by unwinding a string attached at its other end to my starting point, like Theseus in the maze. It is a "my-velocity" which I measure when I read the speedometer of my car. Such measurements of velocity are of great utility.

It is doubtless desirable in scientific discourse to have a name for this sort of velocity with less personal implications than "my-velocity," and I shall follow Ives in calling it "self-measured" velocity. Every moving object has such a self-measured velocity. To determine it numerically, I merely have to move alongside the object and determine my own self-measured velocity in the standard way. In the absence of any presumptive evidence to the contrary, it is to be assumed that this self-measured velocity is unique. This means that my self-measured velocity is the same whether I carry a Hamilton or a Bulova, or in other words, that any effect of my motion on my timepiece is the same, no matter what the details of its construction, provided only that it is a "clock." The effect of the motion on the rate of a clock, or indeed how the rate of a moving clock may be defined, is immaterial. Whatever it may be, it does not enter the definition of self-measured velocity, which specifies a unique procedure and therefore a unique result for any given motion.

It is for experience to decide whether this "self-measured" velocity may be usefully employed in describing the physical world. It would appear unavoidable that in a single frame of reference a description in terms of self-measured velocities is significant, because such a description is unique. The present presumption is that to the extent that relativity theory is correct, self-measured velocity is an internally consistent concept in the broader context of systems moving with different velocities. It is certainly a self-consistent concept to terms of the order of magnitude directly perceivable.

Corresponding to "self-measured velocities" there may be "self-measured times," the times measured on a transported clock. If the velocity of transport is uniform, this time is often called the "proper" time in conventional expositions of the theory.

We have thus far recognized two kinds of "velocity": "two-clock velocity" and "self-measured velocity." We may also describe both of these as "one-way" velocities, since they pertain to a single direction of motion. But we also sometimes encounter a different sort of velocity, as in the case of a ball which bounces back when thrown against a wall or that of the echo from a distant building. This sort of velocity is measured by dividing the total distance, from here to the point of reflection and back, by the time of flight. This time is measured here on a single local clock. We may call this a "two-way" velocity; it is also a "one-clock" velocity. This "two-way" velocity, which may be regarded as a special case of a "round-trip" velocity, applies to anything which comes back to its starting point by a closed path which does not necessarily anywhere retrace itself. The round-trip velocity is to be found in the natural way by dividing the total length of the path traversed by the time interval between departure and return as measured on the single local clock at the starting point.

These different sorts of velocity are all the same within inappreciable differences for ordinary objects, but it becomes very important indeed to distinguish them when dealing with the phenomena of light propagation or other phenomena of high velocities.

It cannot be emphasized too strongly that all these different "velocities" have something in common — they are all velocities *relative to* something. We may talk of the velocity of one member of a pair of particles with reference to the other member of the pair, or we may talk of the velocity of a car with respect to the road. Or if we define velocity as change of position in unit time, the change of position has to be with respect to some frame of reference in which position is determined. Velocity *is* a relative concept, and whenever the physicist allows himself to speak of velocity with an implication of absoluteness, he is either forgetting something or is tacitly implying something he has not taken the trouble to make explicit. Sometimes, for example, the Michelson-Morley experiment is described as showing that "absolute" velocity does not "exist." Of course it does not exist, because it is not that sort of thing *by definition.* What the physicist is actually saying here is that there is no evidence for the existence of the old-fashioned ether, which if it existed could be taken as a universal frame with respect to which velocities could be measured.

One of the most insidious, and because it is so insidious, one of the most vicious formulations of this point of view is: "Relativity theory says that if two frames of reference are moving with respect to each other, it is impossible to say which frame is 'really' moving." The usual implication here is that nature is so constructed that it is impossible to make the decision. The impossibility is entirely man-made. This point of view is behind some of the intuitive difficulties exploited in some recent discussions of the paradox of the "space traveler."

Optical Phenomena

Optical phenomena are basic to the whole subject of relativity, and we must subject them to some sort of critical examination. In their entirety they are complex, and "light" requires such parameters as color, intensity, plane of polarization, degree of coherence, and doubtless others to describe it fully. For our purposes, we need consider only certain limited aspects of these phenomena, which can be conveniently grouped into two major classes. First, there are the phenomena particularly characteristic of relativity theory in which we make our descriptions from the point of view of different frames of reference moving relatively with respect to each other. These phenomena we reserve for discussion in a later chapter. Second, there are the phenomena we have to consider in a single frame of reference, which we discuss in this chapter. These phenomena may be divided into two groups. There are the phenomena of more specific concern to relativity, and it turns out that these are almost exclusively concerned with the "propagation" aspects of light. Then there are broad aspects which are prior to the use of even a single frame of reference. Our perceptual world is to a great extent a visual world, and our intellectual tools and whole way of thinking are based to an extent we seldom realize on our visual experiences. We consider this latter aspect of the situation first. We want to be able to answer such questions as: What sort of thing is light anyway? To what extent do ordinary concepts apply to it? In particular, does the concept of velocity apply, or that of causality?

Light, or radiation in general, plays a role at all levels of experience, but the role it plays may be different at different levels. It may be because of this difference of roles that the worlds of relativity theory and quantum theory are so greatly different from the world of everyday life.

The common-sense world of everyday life is composed of external objects whose relation to us and to each other is continually changing and which constitute our most immediate and vital concern. At this level, the role of radiation is to give us information about these external objects. At this level, the radiation which informs us about an object is without reaction back on the object. This information may be of the most varied sorts, including information about the location of objects and such of their properties as can be inferred from general appearance combined with past experience. These properties include shape and color most immediately. The radiation with which we are concerned at this level is entirely in the optical range, and the corresponding phenomena involve the sense of sight. In the realm of the phenomena of sight, it is possible to develop a completely autonomous and independent science of optics, which includes much, but not all, of what is usually understood as geometrical optics and physical optics. The subject of optics which thus arises to deal with the experience of daily life is sharply separated from what is understood by mechanics, because in optics such things as forces, masses, and velocities do not appear. At the level of naïve, first experience, we have no inkling that light has a "velocity" or that a beam of light exerts a pressure. The reason that ordinary experience has no inkling of these things is that our ordinary instruments, including our senses as instruments, are not sensitive enough, since the velocity of light is too high and its pressure is too low for measurement with such tools.

But presently, with our increasing command of the world around us, it has become possible to add both a velocity and a pressure to the other properties of visible light or radiation. Two sorts of extension of ordinary experience are capable of giving us the velocity of light. The velocity can be measured either by extending the distance over which light travels to astronomical magnitudes — as in dealing with the satellites of Jupiter — when we are able to measure the velocity with ordinary timepieces; or by extending the precision of the measurement of short time intervals combined with terrestrial distances, as the ingenious experimenter Fizeau found how to do with his toothed wheel. Similarly, we did not acquire command of the pressure of light until we had devised new methods which permitted the measurement of forces of a smaller order of magnitude than those formerly accessible.

The instrumental advance which gave us command of forces and times of a smaller order of magnitude than those previously controllable, and of distances of a greater order of magnitude, opened up to us a new "level" of experience. At this new level, light (radiation) is acquiring properties which at the ordinary level we call "mechanical." Hence the world is taking on an aspect of greater unity. This is most gratifying. Despite this pleasant result, we should not let down our logical guard and forget the logical necessity, when we encounter new levels of experience, of checking the continued validity of generalizations we had previously made on the basis of a more limited experience. In particular, we have to reassure ourselves that generalizations such as Newton's laws of motion, or the conservation laws, or even more generally the causality principle, continue to hold. A moment's consideration shows that some modification will be necessary in such generalizations as, for example, the conservation of momentum, because the pressure exerted by a ray of light on leaving its source does not find its reacting pressure until after the lapse of the finite time required for the light to be absorbed at the final sink. Similar difficulties are presented by the conservation of energy. If energy is to be conserved, we must recognize energy associated with radiation, in transit in so-called empty space, between source and sink, and with no association with ordinary matter.

But is the association of energy or momentum with the "empty" space between matter something that has instrumental significance? Even if theoretically it has instrumental significance, may it be that in the present state of development of instrumental technique an order of sensitivity might be demanded in the instruments beyond any yet attainable, and that to acquire this degree of sensitivity would demand pushing on into a further new level of experience? And granted that such an association of energy and momentum with space can be given instrumental significance, what is the assurance that the principle of causality will continue to hold?

To what extent does the principle of causality apply to the reception and emission of radiation, anyway? What is necessary to permit the prediction "in principle" of an emission or an absorption? We have to suppose, in the first place, that an emission or an absorption is detectable and also measurable. We may suppose that this is done in mechanical terms, by the kick of momentum when radiation is emitted or absorbed. Because of the kick of radiation, we have to begin to talk at this level about the reaction of the agent of information on the object of information. We are concerned here not with single photons or with emission and absorption by single atoms, but with gulps of radiation containing many photons. We suppose that the sensitiveness of our mechanical instruments has been sufficiently increased to permit measurement of the momentum kicks of these gulps of radiation. Such an extension of instrumental sensitivity constitutes the first level beyond that of common-sense experience. This is the level at which the momentum or energy changes associated with either emission or reception of gulps of radiation are detectable and measurable.

We have to ask whether they can be *predicted* at this level. Reception can obviously be predicted, because it *follows* emission, which can be detected. But emission is different; emission may follow the opening and closing of a shutter on a radiation box or lantern containing black body radiation at an appropriately high temperature. The mechanical mass of this shutter may be made as small as we please, and since the direction of motion of the shutter is at right angles to the direction of the radiation pressure acting on it, the forces and energies involved in opening and closing the shutter may be made as much smaller as we please than the forces and energies associated with the emitted gulp of radiation. Hence, the order of magnitude of the energies of the mechanical effects in the domain in which the opening and closing of the shutter is a causal occurrence is lower than that of the energies associated with the gulp of radiation; and, by hypothesis, it is at a level at which we cannot yet control or measure and therefore cannot predict. On this level, therefore, if we want to maintain, as far as possible, the principle of causality, the best that we can do is to say that future acts of reception of radiation are predictable but not necessarily future acts of emission.

What now is involved in our conclusion that at this level of experience future acts of reception of radiation are predictable but not necessarily acts of emission? We must begin by inquiring what is involved in the concept of a system the future of which is predictable. With ordinary implications, in order to predict the future of a system we need a complete knowledge of its present state. If the future states can be predicted from such complete knowledge of the present state, the system is said to obey the principle of causality. We now must ask what constitutes complete knowledge of present state. If the system is a classical mechanical system, then complete knowledge of present state demands knowledge of position, velocity, and masses of all the particles (if the system is one of mass particles), and also of the forces between the particles, in general as a function of distance of separation or of relative position. But if there is radiation in the system, in the form of traveling gulps of radiation in empty space which will presently be received at one or another of the material points, this specification of present state is not enough. For we have stipulated that we must be able to predict the future acts of absorption of the material particles. And experience shows that a past act of emission leaves no permanent trace on the material particle which emits it, so that we have no possibility from measurements on the particles of inferring the traveling radiation in empty space which will give rise to future absorption.

In order to bring such a system under the operation of the principle of causality, we must make some extension of our definitions. It appears that at least two courses are open. We may say that a material system is a causal system if it is possible to predict completely its future in terms of a specification of its present state and of its states through an interval of past time (in the extreme case, through all past time). Or we can say that the system is to be extended to include not only what we call the material objects but also all points of the space between the material objects. Our original definition of the present state of a material aggregate was "the total of all the instrumental readings it is possible to make now on the material bodies of the system." We extend this definition to say that "the present state of a system is the aggregate of all the instrumental readings that can be made now on the material objects of the system plus readings in the apparently empty space between the objects." Since without further qualification such a system might reach to infinity, we must add the further requirement that the system be contained in some sort of impervious enclosure by which it can be "isolated" from the rest of the universe.

The second way of extending the system, by including empty space in it, corresponds to the usual way of handling the situation. But the first method, projecting the history of the system back into the past, contains another possible method of generalizing the causality principle which is equally acceptable logically and which corresponds to what we actually do when confronted, for example, with plastically strained bodies which admit no reversible displacement whatever. Here the causality principle gets generalized into what I have called the principle of "essential correlation": When the present state and entire past history of a system repeat, all future behavior repeats also. We can make shift logically to get along with such a principle, for we can reduce the world to formal order by means of it, but such a principle gives us no hold on any *mechanism* by which we can *understand* how the future may be determined by past and present together, a mechanism for which primitive intuition has an almost irresistible craving.

In addition to plastically strained bodies, which we handle by the principle of essential correlation by introducing past history, there is another exceedingly important class so handled in daily life, namely biological systems. A biologist can tell what sort of organism a seed will develop into if he knows the history of the seed, although he would be completely unable to make such a prediction if he were allowed only to examine the present state of the seed by all methods now known. Still more do we need a knowledge of past history to predict the future of a *social institution.*

Whichever way we choose to generalize the causality principle, we encounter a dilemma at the level of experience which begins to associate measurable mechanical parameters with radiation. For at this level, emission of radiation has to be treated as uncaused, because it is not predictable. This level is not as deep as the conventional quantum and atomic level, at which the emission of a photon of radiation by an atom is treated as a matter of pure chance. We have already said that the level here is the emission of gulps of finite numbers of photons under the action of light shutters. If at this level we have to treat emission as uncaused, we shall probably have to deal with the situation statistically, in terms of a chance arbitrarily imposed by fiat. It would be interesting to work out in detail the mathematics of this level of experience. Conventional quantum theory jumps over this possible level of experience without recognizing the possibility of its existence. A result to be anticipated from a working out of the mathematics of the new level would be that it would indicate how, when the analysis is pushed to the next lower level of true quantum phenomena, the motion of light shutters and such things acquires causality in terms of mechanisms too small to be detectable at the higher level. If this could be done, one might hope, by analogy, to get some inkling as to how the now purely statistical situation might acquire causality through the action of mechanisms at present concealed. All this, of course, would not be possible if the complete logical cogency of Von Neumann's argument as to the impossibility of concealed mechanisms is granted.

Going back to the suggestion of the possibility of an analysis of the level of experience which we usually overshoot, it is to be recognized that such an analysis will encounter formidable difficulties, which might even prove to be insuperable, because in practice the different levels of experience are not sharply separated. This difficulty appears at the very beginning, for the common-sense level cannot be adequately handled without bringing in some of the factors by which we later deepen the level to include phenomena associated with the velocity of light and the realm of relativity phenomena. The "here" of common-sense experience is a "here" extended in space with the cooperation of light beams. The ordinary concept of the "instantaneous" velocity of a body, here and now, demands for its precise specification in operational terms a determination of *two* positions of the body. One of these may be its position *here*, determinable mechanically by the sense of touch, for example; but the second position is elsewhere, and information about it is acquired with the cooperation of a beam of light, the velocity of which we neglect for immediate purposes. Or if we avoid this dilemma by the use of two clocks in two different places, the same involvement of the beam of light returns in the procedure for setting the two clocks to synchronism. The difficulty of sharply separating the two levels of experience appears even more formidable when we reflect that in the human eye we have an instrument for the reception of radiation enormously more sensitive than any instrument operating on mechanical principles and enormously more sensitive than any instrument, human or mechanical, for detecting emission.

The causal asymmetry we have recognized between emission and reception has a rough parallel in the world of common sense. We live in a sea of radiation through which we acquire our knowledge of the external world, and this sea of radiation we have come to take for granted without ever thinking of it as something which in principle is causally controllable. The world of our experience is a world of emitted radiation, which is always there, waiting for us, and it is not obvious that we can do anything about it. The reception of radiation is something quite different; we can make it occur or not at will merely by opening or closing our eyes. It is easy to imagine that the nature of our perceptions and our concept of the world would be entirely different if we had to take the same active part in procuring the emission through which we become aware of the world as in procuring the reception through which we become aware of it. A race of beings living in total darkness and becoming aware of their surroundings through the use of optical probes, such as deliberately manipulated hand flashlights, would doubtless conceptualize the world in terms totally different from ours.

Propagation Aspects of Light

We turn now to a consideration of the group of optical phenomena in a single frame of reference, the phenomena of propagation with which relativity is primarily concerned. Relativity theory, as usually presented, visualizes an optical disturbance as initiated at some center and traveling through the surrounding space as an identifiable entity, like a water wave on the surface of a pond. I have emphasized in some of my writings that this method of visualizing the situation assumes more than is necessary. We have no experimental evidence of light as a "thing traveling," but the light of experience is exhaustively described (except under such abnormal physiological conditions as when we receive a blow on the eyeball) in terms of *things lighted.* The nature of our entire perceptual world is determined by this fact. We see the "things lighted" as self-contained *sources of light*, and we are never conscious of the act of reception in our retinas, without which the source would be impotent. We have here the causal asymmetry at a certain level of which we have already been speaking.

But although our perceptions do not usually directly reveal it to us, there still is something progressive about a beam of light. We can make the path of a beam visible with motes of dust in the beam, and when the beam is turned on we can see the head of the illuminated region progressing, if we are sufficiently far away and at right angles to the path of the beam. Now it would seem to make very little difference for most purposes whether we think of the propagation of light in terms of the progress of the head of an illuminated region or in terms of the motion of some more material thing. My present feeling, therefore, is that the device of conventional relativity theory of thinking of light as a *thing traveling* has been innocent and has led to no false result. However, it is just as easy to think of light propagation in terms of the advance of a lighted region, and I prefer to think in these terms.

It is to be specially noted that all the preceding discussion deals with light as a *macroscopic* phenomenon. The intensity of our beams is great enough so that we can siphon off part to illuminate the dust motes with no appreciable effect on what is left. When it comes to light of such low intensity that we have only single photons, this conceptual approach loses its validity. Indeed, whether it is proper to think of single photons in terms of *things* is questionable. Special theory, as a whole, is a macroscopic theory. It is rather paradoxical that the present consensus of theoretical physicists seems to be that it is just in the *microscopic* domain of nuclear phenomena that relativity theory has its most important applications and that we are most sure of the results.

We have everyday experience of at least three other sorts of progressive phenomena. We have ballistic phenomena, such as a projectile. We have transmission phenomena in a medium, such as a wave on the surface of a pond or a sound wave. These two are "causal" phenomena. The third sort of progressive phenomenon is typified by the patch of illumination on the clouds produced by a sweeping searchlight. The progressive aspect of what we see here is "noncausal" in the sense that the successive illuminations are not causally connected, for we could suppress at will any part of the illuminated track by suitable screens with no effect on the rest of the track. The progressiveness of an ordinary beam of light, on the other hand, is "causal" for reasons that will appear shortly.

There is, then, something progressive about light — we can see it progressing. However, just what we see depends vitally on our position with respect to the beam. If we are in the track of the beam, we may see the whole track of the beam flash into illumination simultaneously, or we may see the illuminated region receding from us at a definite, measurable rate. If we attempt to measure numerically what we see, and do this naïvely with the readings on our local clock as we obtain them without correction, we shall say that the velocity of the beam is either infinite or $c/2$. And if the beam is at a great distance at right angles to our line of sight, the velocity we directly see is c. But now we require correction of these different naïve velocities we see in order to get the "correct," "real" velocity. How shall we make the correction, and which one of the possible kinds of velocity we have been discussing do we wish to obtain after we have made the correction? I think our intuitive impulse would be to say that we want the "one-way" velocity. And, having obtained our one-way velocity, we would expect to be able to answer various factual questions about the propagation of light, especially what the numerical value of this velocity is and whether light is propagated isotropically — that is, whether the one-way velocity of light is the same in all directions.

Before we try to answer this question, one other comment will be made about the properties of a beam of light. It has a directional aspect, as distinguished from any "propagational" aspect. Propagation phenomena imply some sort of temporal aspect. Light, however, has a directional aspect entirely apart from even such qualitative temporal aspects as *before* and *after*. We observe that only one side of the dust motes is illuminated, so that there is some sort of spatial asymmetry. The spatial asymmetry may be shown also by interposing an opaque screen across the whole path of the beam. If we interpose the screen at one end of the track, the entire beam is suppressed; if we interpose it at the other, nothing happens. The beam of light thus not only has a track but also has a direction in the track. This purely geometrical asymmetry in the beam will later be connected with the idea held by many writers that a beam of light determines a causal sequence and, through the medium of causality, determines a necessary order in time. (The concept of cause is here taken to imply that an event A cannot be the "cause" of an event B unless it is earlier in time.) The geometrical asymmetry in a beam of light is also revealed by the mechanical pressure it exerts.

Nature of the Velocity of Light

We return now to the question of measuring the velocity of light. It is universally agreed that the velocity of light is one of the most important parameters of nature, and a great deal of experimental effort has been devoted to determining it. What sort of velocity is it which has thus been determined? The overwhelming consensus seems to be that it is the "two-way," or "go-and-come," velocity which has been measured in every experiment. This obviously applies to

Fizeau's toothed wheel and to the Michelson-Morley experiment (light travels back and forth along each arm of the interferometer), but it also applies to such measurements of high accuracy as the recent ones of Townes and Cedarholm at Columbia. For the latter make use of standing resonance waves, and a standing wave involves two-way propagation. The only apparent important exception is the classical astronomical method in terms of the retardation of the time of the eclipses of the satellites of Jupiter when observed from different points in the earth's orbit. This method, recently called to renewed attention in this connection by Dingle, ostensibly gives the "one-way" velocity across the diameter of the earth's orbit. However, questions which will be considered later arise in connection with the logical interpretation of these measurements, and the situation is not as straightforward as it might seem.

The obvious question for us now is whether some new experimental method could be devised which would give the true one-way velocity and thus permit an answer to the question, among others, as to whether the propagation of light in empty space is "really" isotropic. For instance, is it not now possible, with greatly improved experimental techniques, to time the passage of a distant light beam at right angles to the line of sight and find by direct observation whether the velocity from right to left is the same as that from left to right? One reason why this sort of experiment has not been tried hitherto is that it has been too difficult technically. The point I wish to make here is that there is a much more important objection to trying to perform such an experiment: any results it might give would be completely irrelevant. It must be admitted at once that it is perfectly possible, "in principle," to measure such a one-way velocity and to obtain a unique result. In this sense "one-way" velocity has physical "reality." But the reason it is irrelevant for our present purposes is that what we get by such a measurement is information about the combination "properties-of-light-plus-the-method-by-which-time-has been-spread-through-space." Since the way in which time may be spread through space is variable independently of any properties of light, we cannot, from a measurement of the combination, isolate any information about the properties of light only.

Does this mean that the concept "isotropic light propagation" has no real physical content? Is there not something "objective" connected with the propagation of light, independent of the way we spread time through space? The answer is that there is indeed something here with real physical content and that our physical intuition was correct in thinking that there must be such a thing. The trouble is that we have formulated what we were hunting for in improper terms. In what other terms, then, might we make our formulation? It seems reasonable that we should attempt to make it in terms of things with "real physical content." Now any one-clock datum is such a thing, at least to the extent that it will not depend on the way we spread time over space. Therefore, to the extent that a "clock" is a well-defined physical instrument, we might anticipate that we should have asked our question about isotropy in terms of "two-way" or "go-and-come" velocities, which are one-clock velocities, instead of in terms of one-way velocities, which are two-clock velocities. This does indeed prove to be the case when we work out the details.

Our intuitive demand for a precise formulation of what isotropy is would seem to demand at least that the two-way velocity be the same in every direction, and this is a requirement that can be subjected to direct experimental check. A little consideration will show, however, that this does not cover everything that is required.

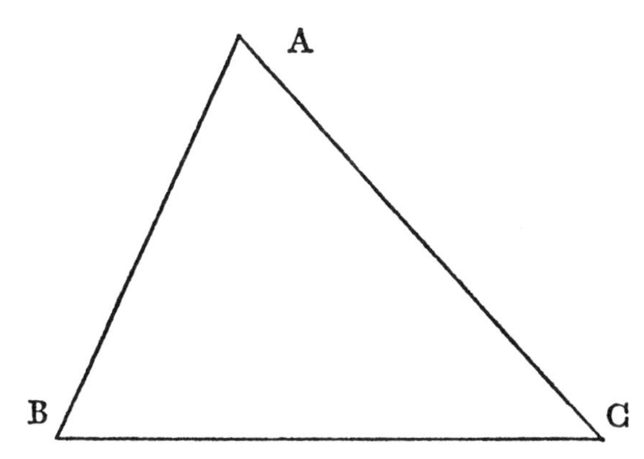

Consider the triangle ABC with clocks at the corners. The clocks are, as always, running at the same rate, but the settings are arbitrary. If we denote the time interval, as measured by the clocks, for light to travel from A to B by (AB) and the corresponding interval for light in the opposite direction by (BA), with a corresponding notation for the intervals characterizing the one-way passage of light along the other sides, then the constancy of the three two-way velocities gives these equations:

$$(AB) + (BA) = 2\left(\frac{AB}{c}\right)$$

$$(BC) + (CB) = 2\left(\frac{BC}{c}\right)$$

$$(CA) + (AC) = 2\left(\frac{AC}{c}\right)$$

Here AB, etc., denote the geometrical length of the sides. Notice, incidentally, that these equations ensure that the two-way velocity in any compound path built up from paths along the sides is also c. For example, the two-way velocity in the path $ABCBA$ is obviously also equal to c.

What now about the round-trip velocity, completely around the triangle? Since we return to the starting point, this is also a one-clock velocity and is therefore independent of the way we have spread time through space. I think the ordinary implications of "isotropic" would demand that this round-trip velocity also be equal to c, and we now impose the corresponding additional restriction. That is, we impose the additional equation:

$$(AB) + (BC) + (CA) = \frac{AB + BC + CA}{c}.$$

The three equations above do not demand this condition; they demand only that if the round-trip velocity clockwise exceeds c, the round trip counterclockwise be less than c by a corresponding amount. To check whether our new condition of isotropy holds, it is therefore necessary to make the measurement only for one direction of passage.

Full isotropy thus demands at least that the four conditions written down be satisfied. Mathematically it is possible that the three two-way conditions be satisfied but not the round-trip condition. If we should find experimentally that this is the state of affairs, we would probably try to associate the round-trip anisotropy with some sort of rotation.

It now appears more clearly why any determination of a two-clock, one-way velocity is physically irrelevant: the settings of the clocks at A, B, and C may be changed in any arbitrary way, with no change in any of the four equations of condition above. The obvious reason is that any changes of setting always occur in pairs in the equations and with opposite signs, and therefore cancel.

It is possible to associate another one-clock velocity with the triangle *ABC*. Thus, we might measure on the clock at *A* the interval required by light to travel from *B* to *C* as we see it. This is also a one-way velocity. We could call this a "*distant* one-way, one-clock velocity." It is for experiment to decide whether this velocity is the same for light traveling from *C* to *B* as for light from *B* to *C*. If we observe this "velocity" for one direction of passage, and if the three conditions on the three two-way velocities along the sides of the triangle hold, then the equations determine the velocity for the other direction of passage. There is, however, no necessary connection between this sort of velocity and round-trip velocity, and therefore no connection with full isotropy. In general, for triangles with arbitrary angles, the "distant one-way, one-clock velocity" will not be the same for the two directions of passage, since one direction will involve a component on the average toward *A* and the other direction a component away from *A*.

If we allow the triangle *ABC* to become isosceles, the equal angles being at *B* and *C*, and allow the vertex *A* to recede indefinitely, we have the state of affairs described earlier when we were asking whether new sorts of experiment were not conceivable that would give the "true" one-way velocity. The analysis of the last paragraph shows in detail why any such measurement is irrelevant. If one still has an impulse to ask what is the "true," or "real," one-way velocity of light, the answer must be that this is an illegitimate question. It is illegitimate because it conceals a self-contradiction. For, by implication, the question is asking what the velocity of light would be if we could eliminate the effect of the way we have chosen to spread time over space, whereas *by definition* a one-way velocity is a two-clock velocity, which by definition involves the way time is spread over space.

Imagine now that we have set ourselves the problem of finding whether the propagation of light is in fact isotropic, and have established by actual measurement that the two-way velocity is *c* in half a dozen different directions chosen at random. Will this satisfy us, or must we keep on trying other directions; and, if so, how many other directions must we try and how should they be selected? Experience with other sorts of progressive phenomena, such as the propagation of a water wave on the surface of a pond, suggests that propagation velocities in different directions are connected and that observation in a few directions should permit calculation for other directions. Is there something corresponding for light? If propagation is controlled by a differential equation, and if, for example, the equation is linear of the second order (which physically means additive effects) and there are no explicit functions of the coordinates or the time (corresponding to independence of the effects of absolute position), then with three spatial dimensions there are eight undetermined coefficients; and one suspects that, at the very worst, determinations of the velocity in eight arbitrary directions would fix the velocity in all other directions.

Relativity theory, as it is conventionally expounded, is not very articulate as to precisely what is necessary here. Sometimes, apparently, all that is assumed is that light has a definite velocity in every direction. And sometimes, as in some derivations of the Lorentz equations, it is assumed that propagation is governed by the conventional wave equation.

As long as it is assumed that the propagation of light is characterized *only* by its two-way velocity in any direction, it would appear that an experimental check of isotropy is necessary for *every* direction. For one can imagine a medium constituted of infinitely fine threads radiating in every direction, each thread having its own characteristic velocity of propagation. Such a medium would permit discontinuous variations of velocity with direction, and verification of the velocity for every direction would appear to be necessary. Hence it would logically appear that, in order to verify the physical correctness of relativity theory, the velocity of light should be checked for every direction. Practically, we doubtless will be satisfied with a single triangle, *chosen at random.*

There is one important proviso to our contention that a one-way velocity, being a two-clock velocity, contains an arbitrary element, in that the way in which time is spread over space is in our control. It is true that when we *measure* a one-way velocity we encounter an arbitrary factor which makes the measure we obtain irrelevant for many purposes. But, paradoxically, there is a *qualitative* aspect to one-way velocity which does not contain an arbitrary element and which therefore has "absolute" significance. We can always decide whether the one-way velocity of some specific object or process is greater or less than that of some other without any convention as to how the distant clock is set. We merely have to arrange matters so that the initiation here of the two processes whose velocity we are comparing occur at the same instant of time on our local clock, and observe at the distant station, on the clock at that station, which disturbance reaches there first. The one-way velocity of the disturbance which reaches there first is the greater. In particular, the statement that no material object can be given a one-way velocity greater than the velocity of light has "real" physical content in that it remains true no matter how we spread time over space.

Perhaps it is this "absolute" *qualitative* significance of one-way velocity that accounts for our obstinacy in trying to find some physical significance in a *measured* one-way, two-clock velocity. However, the possibility of making a greater-less comparison of two one-way velocities depends only on the uniqueness of the time order of events at any local station. The possibility of such uniqueness is concealed in the specification of what is a "clock," which, in order to be acceptable, must be such as to give this uniqueness. We are not saying anything new physically when we say that it is possible to compare qualitatively two one-way velocities.

It is a cardinal postulate of relativity theory that the velocity of light is a constant, independent of the velocity of its source. This postulate applies in the single frame of reference, which is the background of all our present discussion; and it applies to the two-way velocity, not the one-way velocity, as uncritical intuitive thinking is likely to assume. Intuitive thinking seems to find a one-way velocity independent of the velocity of the source acceptable only in the context of a medium in which light is propagated. However, it is recognized that the assumption of such a medium leads to just those difficulties which gave rise to relativity theory. But without the medium there are also difficulties, for independence of the source seems to attach a significance to the *place* at which the disturbance originated, and what is there to give significance to a place in empty space? But with the observation that we are talking about the two-way velocity, this difficulty is somewhat mitigated, although not entirely disposed of. Consider two measurements of the two-way velocity of light between A and a distant station B, at which it is reflected. The first measurement shall use a source at A which is traveling toward B with

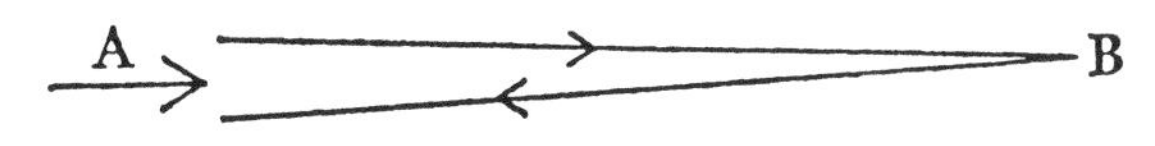

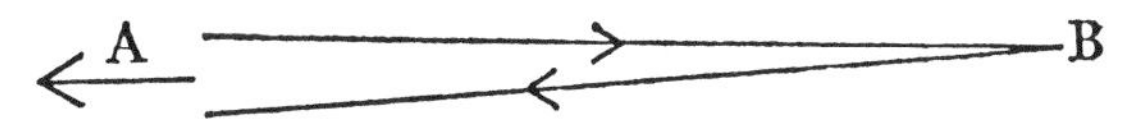

velocity v, and the second measurement shall use a source at A which is traveling away from B with the same velocity. Now in the two measurements there is a certain correspondence between the legs of the light paths, but this correspondence occurs in different order in the two measurements. In the first, light is traveling with the source on the first leg and against it on the second leg; whereas in the second measurement, light travels against the source on the first leg and with it on the second. In terms of relative velocities, the relative velocity of light and source is, in the first measurement, $c - v$ on the first leg and $c + v$ on the second, whereas in the second measurement the relative velocity is $c + v$ on the first leg and $c - v$ on the second. It is not intuitively repulsive that the over-all result should be the same in the two cases.

Practical Limitations on Spreading Time

Despite the fact that we can spread time over space in a completely arbitrary manner, with no effect on any statement of "purely physical" content, in practice we nevertheless do impose very definite restrictions. What is the nature of these restrictions, and what are we trying to accomplish by imposing them? In the first place, we shall not go out of our way to make things needlessly complicated when we set the distant clocks. We have a certain amount of physical experience preliminary to spreading time over space. We have a certain amount of knowledge, for example, about the way clocks behave when transported to a distance. We know that we can travel to a distance without any noticeable effect on the watch in our pocket. It continues to keep time after we arrive. We have already used this property of transported clocks in our stipulation that the rates of clocks be the same, no matter where situated. But we also know that when we move about we do not need to change the setting of our watch (short of crossing to another standard time belt). For terrestrial distances and within everyday limits of accuracy, there is no perceptible difference between the various concepts we have taken so much pains to distinguish logically. Such things as one-way, two-clock velocity, one-way, self-measured velocity, and two-way, one-clock velocity all are indistinguishable, as are also any differences between the corrections we may apply to our clocks when allowing for light velocity or transporting them. Now any method of spreading time over space we will be willing to accept must be such that it will preserve all these approximate equalities. This means that the arbitrariness we will tolerate in our setting of distant clocks must all be within the limits of time required by light to travel from one clock to another. In particular, this means that the maximum arbitrariness in setting clocks anywhere on this earth be not more than about one-fifteenth of a second.

It is characteristic of relativity theory that in giving a

specific answer to the question of how best to spread time its paramount concern is with the propagation properties of light. This fact about relativity theory is often put in a spurious historical perspective, since one of the earliest applications and arguments for the correctness of the theory was afforded by the Michelson-Morley experiment. Unfortunately for this easy idea, it seems that Einstein was influenced to a very slight degree, if at all, by consideration of the Michelson-Morley experiment. At least, this is what Einstein himself said in his later years. He was, however, much concerned with the propagation of electromagnetic disturbances and with the field equations of electrodynamics. Since light is an electromagnetic disturbance, the fact remains that one of the chief concerns of relativity theory has been with the propagation phenomena of light.

What restrictions, now, is this concern with the propagation of light going to impose on the way we set a distant clock? Examination of what physicists have done and said in this situation shows that no method has any chance of acceptance if it violates the following cardinal demand: If I despatch a beam of light to a distant clock, then the time at which I see it arrive at the clock must not be earlier than the time on my clock here where I despatched the beam. Why does everyone insist on this requirement? I do not believe that any logically cogent justification can be found for it. Operationally, time *here* is different from time *there*, and there is no intrinsic reason why the two-clock interval for light to get from here to there should not be negative. But if one does admit such negative time intervals, one will experience a psychological wrench in passing from thinking about events all taking place here to events happening in different places. This wrench is so severe and disagreeable that one tries to avoid it if possible. Just why is such a wrench so upsetting? The answer that the conventional relativist would give to this question, if indeed he would bother with the question at all, is that there is a tie-in with our ideas of causality, and philosophy has taught us to believe that there is something pretty fundamental and universal about causality.

This is not the place for an elaborate analysis of the concept of causality, which in its entirety is most complicated, as witnessed, for example, by Mario Bunge's* recent book on causality. As ordinarily understood, on the plane of daily life and when we are dealing with events all in the same locality, a causal connection between two events implies an ordering in time. If event B is the result of event A (which is another way of saying that A is the cause of B), then B follows A in time. But in addition to complexes of events all in the same place, we also encounter chains of events in different places which are connected in such a way that the concept of causality applies to them. Thus the arrival of a rifle bullet at a distant target is the "effect" of the discharge of the bullet from the rifle here. Now if we are going to maintain that a causal connection is inextricably tied up in some way with an order in time, and if we regard a causal connection as something intrinsic or "absolute," then we will demand of any method of spreading time that it be such as to preserve the time order of events spatially separated but causally connected. This means that if B is the result of A, the time of event B on the local clock where event B occurs must be later than the local time of the event A on the clock where A occurred. As we have seen, this is pure psychology, a demand that we make in the interests of simpler thinking.

* M. Bunge, *Causality*. Cambridge: Harvard University Press, 1959.

Now if anything is obvious physically it is that the propagation of light is a "causal" phenomenon. We can arrange a signaling code in accordance with which we can make the occurrence of event B there follow the occurrence of event A here. The complete mechanism of the causal tie-in of A and B can be made automatic if desired. We shall therefore demand that we set the distant clock in such a way that the time at which a light signal arrives at the distant clock, as measured on that clock, shall always be later than the time at which we despatched the signal, as measured on our clock. This single requirement still allows indefinite latitude in the way we set the distant clock. There is, however, a further very definite restriction if we impose the additional requirement of symmetry, namely that the time at which a light signal leaves the distant station for our local station must always be earlier than the time at which it reaches us (the two times being measured on the respective local clocks). With this requirement, the latitude in setting the distant clock is narrowed down to the two-way interval for light to go and return between the stations.

Simultaneity

The concept of simultaneity is obviously connected with this question of setting distant clocks. Local simultaneity, we assume, is a concept not in need of analysis; but distant simultaneity — that is, simultaneity between two events occurring at a distance from one another — is a more complicated thing. By definition, two distant events are simultaneous if the times on the two respective local clocks at which the events occur are the same.

Reichenbach* has shown in an analysis of distant simul-

* H. Reichenbach, *The Philosophy of Space and Time* (New York: Dover Publications, 1958), §19: pp. 123–129.

taneity extensively quoted by Grünbaum* and others that the time order of distant events causally connected by light signals is maintained if the distant clock is set anywhere within the latitude specified above. Specifically, Reichenbach's equation for setting a distant clock is:

$$t_2 = t_1 + \epsilon(t_3 - t_1),$$

where t_2 is the corrected distant local time (corrected after the clock has been set) at which the light signal arrives at the distant clock, and t_1 and t_3 are the respective times on the local clock at which the light signal leaves and returns by reflection. The demands for a causal propagation — that is, for an invariable order in time — are met if ϵ is given any value between 0 and 1.

Unless ϵ has the value $\frac{1}{2}$, the state of affairs described by Reichenbach's equation is unsymmetrical in that the one-way, two-clock velocity of light from here to the distant clock is different from the corresponding velocity from the distant clock back to here. Einstein adopted the value $\frac{1}{2}$ for ϵ, thereby ensuring symmetry in the one-way velocity of light. Einstein knew perfectly well when he adopted this value for ϵ that he was not saying anything "real" about light by itself, but was only saying something about light and the way he was spreading time through space. For Einstein repeatedly emphasized that the choice of the value $\frac{1}{2}$ for ϵ constituted a "definition" of distant simultaneity. In other words, in this context it is a "convention" to say that light is propagated with symmetrical, one-way velocities.

This word "convention" has stuck in the crop of many

* A. Grünbaum, "Logical and Philosophical Foundations of the Special Theory of Relativity," in A. Danto and S. Morgenbesser, eds., *Philosophy of Science* (New York: Meridian Books, 1960), pp. 401–412.

physicists, who have made strenuous efforts to avoid it and with it the implications of physical unreality they find. One of the most elaborate of these attempts was made by Herbert Ives. In all these various attempts there has been much discussion of such things as the nature of simultaneity and other possible methods of setting distant clocks. Much of this discussion I believe has been beside the point and has involved misconceptions. Part of the discussion has been concerned with the transfer from one frame of reference to another moving with respect to it. We are not yet ready for this aspect of the discussion, but there are considerations with regard to setting clocks and distant simultaneity in single frames of reference we may now profitably examine.

A good part of the discussion has been concerned with trying to find some other method of setting distant clocks, one which would not make use of light signals and which conceivably might give a hold on the "absolute" simultaneity about which Newton and the physicists of his era used to talk. In this search for other methods of setting clocks it usually seems to be assumed that any acceptable method must involve in some way a causal propagation, and often the argument is made to depend on the postulate that there is no causal propagation with a velocity greater than that of light. The various discussions, taken in their entirety, manifest a vague and complex tie-in between various concepts, which may be illustrated by the following statements culled or paraphrased from various authors.

> A distant event is simultaneous with an event here if it occurs at the same time. In order that I can decide whether it is happening at the same time, there has to be some method by which news of the event at a distance can be transmitted to me. In transmitting news of the distant

event some sort of "signal" has to be used. This signal involves a causal process of some sort because by means of the signal I could couple together events here and at a distance. The margin of uncertainty in my decision as to whether the distant event is simultaneous with the local event is connected with the velocity with which the signal is propagated. If the signal could be propagated with infinite velocity there would be no uncertainty, and I could use an infinitely fast signal to thus determine an "absolute" simultaneity at a distance. But no causal effect can be propagated faster than light, and hence no signal exists faster than light. Finally, since distant clocks *can* be set by the use of signals from here to there, any method of setting distant clocks must involve the use of signals, and therefore, in the most favorable case, the use of light signals.

It will be seen that some of these statements are vague, and that the various points of view are by no means logically connected or even logical by themselves. I believe, however, that one or another of these statements represents points of view which play a part in most if not all the accepted argumentation about relativity theory.

What shall we demand, in general, of any acceptable method of setting distant clocks, or spreading time through space, or defining distant simultaneity, these three things being roughly equivalent? In the first place, if the method is one we can apply ourselves and is not merely something the results of which are to be taken as "given" in the mathematical sense, the method must be describable. If one examines the specification of any physical process whatever, it will be seen that essential parts of the description have to be given in the context of some sort of reference system. If more than one reference system is conceivable, there has to be some method of picking out the particular system being used from all other possible systems. For present purposes we do not need to inquire further how we pick out and specify some particular system, but we shall merely assume that it has been done (perhaps by exhibition if no other method presents itself) and that our conversation is in the context of a single frame of reference. This single frame here constitutes the "stationary" frame assumed in the discussion of this chapter. Given now the frame, the physicist would further insist that any acceptable method of setting distant clocks be unique. Given only that it is unique and describable, so that the physicist can apply the method for himself and verify any ostensible state of affairs, he would make shift to get along with *any* method of setting distant clocks and would not be defeated even if the method did not meet the practical requirements we have already discussed. He might expect to find the interrelations of the readings on the various clocks terribly complicated, but if only he knows what the method of setting is, he can manage to get along. But uniqueness he must have. I believe that in this context the connotations of "uniqueness" are often confused with the connotations of "absoluteness." An absolute method of setting clocks is also unique; the converse is not always true. All we need or use in the context of a single frame is uniqueness.

Distant clocks may be set uniquely by the use of appropriate signals, but there are other methods which do not involve the use of signals or of causal processes. For instance, methods are possible using superlight velocities. For there are occurrences in nature traveling with superlight velocities, although it is sometimes mistakenly stated that relativity theory does not permit this kind of thing. Perhaps the simplest example is the patch of illumination on the clouds thrown by a distant, sweeping searchlight. If the clouds are far enough away and the searchlight sweeps rapidly enough, the patch of illumination may travel across the clouds with any velocity we please greater than the velocity of light, and furthermore we can see it traveling with superlight velocity. A unique numerical value can be given to this velocity in terms of the distance from the searchlight to the cloud and the angular velocity of rotation of the light, as measured on a local clock at the searchlight. Although there is no upper limit to the velocity obtainable in this way, we cannot actually reach an infinite velocity, for this would demand an infinite angular velocity of rotation of the searchlight, which in turn would demand infinite energies. Given the searchlight, we now set our distant clock as follows. We arrange the searchlight so as to sweep over the track between us and the distant clock. We arrange with an assistant at the searchlight to send us a number of different beams with different prearranged velocities of sweep. We set up a corresponding number of clocks at the distant station. We receive the first sweep on the first clock, which we thereupon set so that the time it records when the sweep reaches it is the same as the time on our clock when the sweep reached us. We do the same for the second sweep with a second clock. Eventually we obtain a succession of clocks set with different velocities of sweep and presumably not agreeing with each other. Now at some later instant of time we read all the distant clocks simultaneously (this involves only local simultaneity and can be done unambiguously). We plot the simultaneous readings of all these clocks against the corresponding velocity of sweep and extrapolate the results to infinite velocity. It is physically obvious that this extrapolation is unambiguous. We then recorrect all the clocks so as to be consistent with this extrapolated reading. This is the setting of the distant clock for infinite velocity of sweep.

According to one of the vague points of view suggested above, the distant simultaneity defined above would be called "absolute" simultaneity. I think the uncritical impulse to use the word "absolute" in this connection can be given a plausible explanation. A method of setting distant clocks is "absolute" if it gives the same result in all systems of reference. The implication now is that a method which uses infinite velocities is absolute because a velocity infinite in one system of reference must also be infinite in all other systems. It is this latter unconscious assumption that is incorrect, because as can be seen from simple inspection of the Lorentz equations, an infinite velocity of sweep in one frame of reference is not infinite in another moving with uniform velocity with respect to it.

The method of setting clocks described above thus cannot properly be described as "absolute," but is nevertheless definite and unique, and there is therefore no reason why it should not be acceptable to the physicist. Furthermore, there is no reason in logic why distant simultaneity defined in this way should not be identical with distant simultaneity as defined by Einstein. In fact, the present presumption is that the two are identical. It is ultimately a question for experiment to decide.

Despite the fact that the distant simultaneity which can be specified with infinite velocities is definite and unique, such a use of infinite velocities has been rejected as illegitimate by a number of writers, in particular by Reichenbach, who has given an elaborate analysis to show how such a procedure might work out in detail (pp. 91–97 in *Axiomatik der relativistischen Raum-Zeit-Lehre,* Friedr. Vieweg, Braunschweig, 1924). His objections revolve around the consideration that a superlight sweep cannot be a causal sequence. This may at once be conceded as obvious. For we can suppress reception of the sweeping beam here, by interposing a screen, without at all suppressing its reception at a distance, something we could not do if its reception at a distance were causally connected with its reception here. Assumptions, such as Reichenbach's, of the necessity for a causal process in setting distant clocks can be peremptorily disproved by actual exhibition, as above, of a method of setting distant clocks not involving causal propagation.

Causality and Time Order

Reichenbach's conception of causal relatedness seems tied in with invariable order in time. This is in accord with classical ideas for events all in the same locality; it is carried over to events separated by a distance in the demand that when B is the result of A, which occurs at a distance, any acceptable method of spreading time over space must be such that the time at which B occurs on its local clock must be later than the time at which A occurs on its local clock. We have seen that in accordance with this criterion the value of ϵ in the equation on page 56 is restricted to positive values less than unity. Now this is a perfectly possible restriction, and furthermore it is a natural one if it is assumed that the time order of separated events is in some way intrinsically connected with causality. It appears to me, however, that this idea of a necessary invariable time order of causally connected separated events has no justification from the point of view of pure logic. The very concept of a time order of spatially separated events has meaning only in a context in which time has already been spread over space, and this requires independent specification. Furthermore, I very much question whether the concept of causality is necessarily connected at all with the concept of order in time. We may so connect it if we desire — Reichenbach, among others, has shown how to do it, and it is doubtless convenient so to do because it involves a minimum wrench with established methods of thought. But I question whether there is any necessary connection between the two concepts. Already physicists had recognized that for purely local events the concept of causal connection as determined by invariable order in time was not a fruitful one. The idea of one event as the cause of another and therefore preceding it in time has been replaced by the concept of a whole complex of local phenomena invariably interrelated, not in time, but through the medium of a partial differential equation. The idea of a temporal involvement of causally connected events lingers on, however, when the events are at a distance from each other. But it is to be seriously questioned whether this involvement with time is necessary, and whether there are not purely spatial relations between the phenomena that can be correlated with the ordinary notions of causality.

Consider, for example, a beam of light along the straight line from A to B. We have seen that there are purely geometrical asymmetries connected with this beam, which can be described with no reference whatever to time. The two ends of the track of the beam are different — if we interpose a screen at one end of the track, we suppress the whole beam, whereas if we interpose it at the other end, there is no result. That is, there is associated with the beam a *direction*, which is a purely geometrical parameter in the sense that it does not involve the notion of velocity or progressiveness or any other connection with notions involving time. It is the possession of this sort of geometrical parameter that we are to associate with a causal sequence. We say that light involves a causal sequence because it is possible to associate a direction with it. By the same token, the sweeping patch of illumination of the distant searchlight does not represent a causal sequence because the sort of operation that establishes an asymmetry for the beam of light does not establish an asymmetry for it.

Going still further, I cannot see that the conventional concepts of causality have any necessary connection with the unfolding of relativity theory, although historically the connection is there for all to see. What we have physically at the basis of the theory, as far as light is concerned, is that the velocity of light is independent of the velocity of its source and that no material object is ever encountered moving with velocity greater than that of light. These statements have direct experimental meaning and have been established with high accuracy. They have no necessary connection with causality. This all applies in our single frame of reference. When we deal with two frames together, other considerations enter which we will examine later.

Setting of Clocks by Transport

Our point of view about setting distant clocks may be further emphasized by considering another method of setting them which does not involve causal propagation. We can set our clocks at a distance by transporting them from here. This process has also been considered by Reichenbach,* and his

* H. Reichenbach, *Axiomatik der relativistischen Raum-Zeite-Lehre*, §23, pp. 76–85; also H. Reichenbach, *The Philosophy of Space and Time*, §20, pp. 133–135.

considerations have been extensively quoted and adopted with approval by Grünbaum.* These authors reject this method because it does not give at once a unique result, for the clock transported to a distance by one route and at one velocity does not agree with another transported at a different velocity and by a different route. This, I believe, is an inadequate reason for rejection. For although, as specified above, the naïvely transported clock does not give a unique result, a simple modification can be made that does give uniqueness. We specify, in the first place, that the clocks be transported from here to a distance by the single direct route. Along this route we shall transport the clocks with different specified velocities. These are to be the "self-measured" velocities, which are uniquely determinable without further ado. When we have transported all our clocks, we shall presumably find that the various readings do not agree. We take the difference between the readings of the various clocks against some selected clock, perhaps the one transported with least self-measured velocity, and plot the difference of these readings against the self-measured velocity of the corresponding clock. We extrapolate these differences to zero self-measured velocity, and we set all the clocks so as to agree with this extrapolated difference. By definition, we have now set our clocks for zero velocity of transport. Such a method of setting a distant clock is unique and well defined, involving only actually performable physical operations, and therefore there seems to be no reason why we should not accept it.

* A. Grünbaum, "The Genesis of the Special Theory of Relativity," in H. Feigl and G. Maxwell, eds., *Current Issues in the Philosophy of Science* (New York: Holt, Rinehart & Winston, 1961), pp. 44–45; also A. Grünbaum, "The Special Theory of Relativity as a Case Study of the Importance of the Philosophy of Science for the History of Science," *Annali di Matematica*, **57** (1962), pp. 260–261.

Ives concerned himself much with transported clocks and saw the importance of self-measured velocities. However, he rejected the notion of a clock set for zero velocity of transport, for the ostensible reason, which appears to me somewhat quixotic, that zero velocity of transport is a self-defeating concept, because the clock transported with zero velocity never arrives at the distant station, so that the time thus defined never gets established. It seems to me that by using a mathematical, not a physical, limiting process as above, Ives' objections are answered, and that distant simultaneity defined in terms of transported clocks can be made a perfectly acceptable physical concept. Neither causality nor the properties of light is used in this definition.

If the present equations of relativity theory give us a correct picture of the physical state of affairs, we are to anticipate that distant clocks set by the transport method will agree with clocks set by Einstein's method. Either the transported clock or the sweeping searchlight will give a value for Reichenbach's ϵ equal to $\frac{1}{2}$ within at present undetectibly small experimental error. What is the significance of this? Does it mean that there is something "absolute" about the value $\frac{1}{2}$? What becomes of Einstein's insistence that his method for setting distant clocks — that is, choosing the value $\frac{1}{2}$ for ϵ — constituted a "definition" of distant simultaneity? It seems to me that Einstein's remark is by no means invalidated. He was saying, in effect, that any method whatever for setting distant clocks involves an element of definition, and that in choosing the value $\frac{1}{2}$ he was merely adopting a particular one of these methods. And what becomes of Reichenbach's assertion that there is no physical reason for selecting one value of ϵ in preference to another, provided only that it is included between his limits of 0 and 1? What Reichenbach was essentially saying here was that any description of the world in terms of one-clock data will be unaffected by the way time is spread through space — that is, will be unaffected by the precise value of ϵ. But in the light of our analysis, this statement would appear to be trivial and without physical content. Furthermore, in refusing to accept values for ϵ outside the range from 0 to 1, Reichenbach was subscribing to a concept of causality which, although convenient, is not necessary.

Even if the settings of distant clocks defined by the sweeping searchlight or by the clock-transport method agree in giving the value $\frac{1}{2}$ for ϵ, nevertheless the decision to use one or the other method is a decision in our control, involving a corresponding *definition* of distant simultaneity. The fact that these two methods agree in giving a result obtainable also by another method is in no wise a logically necessary fact, but is something that has to be established by independent experiment.

Mathematics and Time

We can see by looking that the mathematics of our single frame of reference is a mathematics with extended time — every x has its associated t. This is almost unavoidably the case. For even if we could devise a mathematics in which the significance of the symbol t was restricted by fiat to times measured on the clock at the origin, we would presently find ourselves having to express in some way the essential fact that the precise location of the origin is a matter of indifference, and that physics expressed in terms of time intervals at an origin arbitrarily displaced in either time or space is the same as the physics of the time intervals at the original origin. And since, by definition, time intervals are differences of times, we should have to extend time in some way to every point.

Since the mathematics is thus a mathematics of extended time, it follows, in general, that the time intervals of this mathematics are two-clock intervals and its velocities are one-way velocities. The physical significance of what the mathematics says about velocities is therefore limited in the way that our discussion has shown the physical significance of any statement about one-way velocities is of necessity limited. It remains true, however, that a physics expressed entirely in terms of one-clock data at any single point is more significant, in the sense that when we use two-clock data we are talking about both "nature" and the way we have elected to spread time over space, whereas when we use only one-clock data we are saying something about "nature" alone. And in the last analysis, since any valid physics must offer a valid description of my own experience, and since my time intervals are of necessity one-clock intervals, it would seem that if we push our analysis far enough, the requirement that the time intervals be one-clock intervals will be automatically met.

We return for a moment to Dingle's suggestion that Bradley's original method of measuring the velocity of light in terms of the observed retardation in time of the eclipses of the satellites of Jupiter is a measurement of a one-way velocity. If our considerations above are correct, it is impossible in principle to measure a one-way velocity by observations on a single clock at a single station. Any apparent one-way velocity obtained in this way must conceal somewhere an assumption about the way time is spread over space. The way such an assumption enters here is evident on a little consideration. We assume that the satellites of Jupiter themselves function as clocks, and that time is spread over space through the functioning of these clocks. This has to assume the validity of the ordinary laws of mechanics at a distance, and experimental proof of this validity would involve a logically circular process.

Mechanics

We turn now from these considerations, mostly connected with the phenomena of light propagation in a single system, to more extensive phenomena, including in particular those of mechanics. In our discussion of elementary mechanics we pass at once to the full complement of three spatial dimensions, because in passing from one to three dimensions, certain issues arise which are not foreshadowed in the one-dimensional situation, so that the one-dimensional analysis does not really constitute a simplifying introduction. We shall, however, make the customary simplifying assumption of beginning with a system of massive particles. Furthermore, we shall assume that all the particles in the system are "force-free."

Force

"Force" has been the subject of extensive and often controversial discussion. In some contexts, "force" has come to be used in a highly generalized sense. In general relativity, we have "forces" arising from space-time curvature. In classical mechanics, many expositions freely use such expressions as "centrifugal force." Such forces, for which the general term "motional force" is often used, arise in connection with non-

Galilean systems of coordinates and are such as would have to be applied to make the motion in a Galilean system the same function of the coordinates that it actually is in the non-Galilean system. In this present exposition I shall conceive of a force in as nearly as possible purely physical terms. I shall not speak or think of such motional forces as a "centrifugal" force or a "Coriolis" force. When a stone rotates about a fixed center to which it is tethered by a string, I shall speak only of a "centripetal" force — that is, the force exerted by the tension in the string. This force is unbalanced and accordingly produces an acceleration in the body on which it acts, an acceleration which results in the curvature of the path. In accordance with this point of view, a "force" is an aspect of the interaction between bodies. There are, in general, two sorts of interaction — interaction between bodies in contact and interaction between bodies at a distance from each other.

"Interaction" is doubtless a loose concept; it finds its meaning in the context of our whole physical experience. Roughly, there is interaction between bodies A and B if the observable behavior of either in the absence of the other is different from the behavior in its presence. To establish a presumption of interaction, it is obvious that observation is necessary over a wide range of conditions. The two kinds of interaction announce themselves in different ways. If it is interaction of bodies in contact, then there are elastic distortions in the regions of contact. A special case of such interaction is the force exerted by a stretched thread. The force may be measured by the amount of stretch. If the bodies are not in contact, then an interaction between them announces itself in the peculiar way in which the distance of separation of the bodies changes with time. The precise way in which we process our observations of the change of distance with time so as to analyze out the interaction is a complicated matter, which involves a knowledge of the principles of mechanics. The mere existence of the subject of mechanics as a logically consistent edifice shows that such an analysis is possible. For our immediate purposes, the most important sort of interaction between separated bodies is gravitational. There are other sorts of distant interaction — electrical, magnetic, and perhaps others — but we shall assume that our experience has been extensive enough so that we can tell when these other sorts of interaction are present and avoid them when we wish. Gravitational interaction becomes rapidly less when the distance between the interacting bodies increases. Our experience enables us to say in any particular case whether the distance is great enough and the interaction small enough to allow us to disregard it for the purpose in hand.

Establishment of a Galilean Frame

Going back now to our system of massive particles, the condition that they be force-free means that no two are in mutual contact, that there be no strings connecting any one with any other, that they are far enough apart so that mutual gravitational interactions may be neglected, and that any external objects are far enough away so that they exert no appreciable gravitational effects. Subject to these restrictions, we consider a swarm of such particles in their most general possible state. At first appearance, the situation is completely chaotic. There are neither fixed points nor fixed directions to which a description can be tied.

Certain physical measurements can be made on the

swarm, however. We assume, in the first place, that the members of the swarm can be identified; the possibility of identification involves perhaps nothing more than the possibility of continuous observation. Given identifiability, we single out some particular pair of particles and determine how the distance of separation of the members of the pair changes with time. The operation of determining the distance between the members of the pair is performable, given only the pair, and is independent of any external frame of reference. The distance might be determined by stretching a tensionless tape between the members, or it might be determined optically by timing a light signal between them. If now we determine the distance of separation of all possible pairs as a function of time, we shall find that, in general, distance changes with time in a most complicated way. In a great many cases, if the members of the pair are fairly close together, the distance of separation will at first diminish and then increase again. We define the "velocity" of pair separation as the time derivative of the distance of separation. In general, this velocity will reverse sign as the particles first approach and then recede. This means that, in general, the pair distance is subject to *acceleration*, since the time derivative of mutual velocity does not vanish. Not only can we observe pair distances, but also at any given instant we can draw lines between different pair members and find the angles between these lines. These angles will also be changing with time in a most complicated way. Although we can determine whether these angles are changing, we cannot yet answer such a simple question as whether any individual massive particle continues to move in the same direction. This is because we have no means of fixing a single direction.

Order begins to emerge from our chaos when we observe

that among the great number of particles with which we might pair any particular individual, some are singled out by the property that their distance to the particular individual is changing without acceleration in time. If we are fortunate, we may find a particle such that when it is paired with three other particles, all three distances are changing without acceleration. When we have found such a particle, we make the remarkable discovery that the three angles between the three lines connecting it with the other three particles do not change with time. If the three lines do not all lie in the same plane — and it would be only by luck if they did — we have here the makings of a frame of reference and a coordinate system. We get a system of oblique coordinates merely by laying out rigid rods from the particle in the three directions. Along these three directions the other members of the respective pairs are receding with uniform velocity. Having the oblique system, we can at once convert it into a conventional orthogonal system by simple changes.

Once we have a coordinate system, a great simplification appears in the motion of all the other particles. Since we now have the means for specifying directions, we can begin talking about vectors. It appears that every particle continues to move in the same direction — that is, continues to move in the same straight line. Furthermore, it continues to move in this straight line with uniform velocity — that is, without acceleration. And since the motion may be represented by a vector, the component of acceleration along each of the three axes is zero.

Finally, we waive our restriction that the particles be force-free, and allow one of them to be acted on by a force, perhaps by pulling on it with a thread. We shall now find that the motion of the particle is in accordance with the

Newtonian law: force equals mass times acceleration (or the corresponding relativity law if we make the measurements accurately enough). Since we are dealing with vectors, this implies that the component of force in any direction, and in particular along one of the axes, is equal to the component of acceleration in the same direction multiplied by the mass. The great advantage of describing motion in terms of components along the axes is apparent. The simplest formulation of the condition that a vector be without acceleration is that its components along fixed directions are without acceleration. If we express the vector in other terms — for example, in terms of a direction and an absolute (scalar) magnitude — the relations become less simple, for both the direction and the scalar magnitude of an unaccelerated vector, taken by themselves, may be and in general are subject to acceleration.

Having obtained our frame of reference with its three axes, we can invert the usual laws of motion and obtain the laws of motion for the particle pairs, which in the first place were all that were accessible to observation. The parameters to be associated with a pair of particles (1) and (2) are: a single parameter for a sort of joint mass, M_{12}; a single parameter for the joint distance, l_{12}, with corresponding pair velocities and pair accelerations dl_{12}/dt and d^2l_{12}/dt^2; a pair kinetic energy $(\text{K.E.})_{12}$, and a pair potential energy $(\text{P.E.})_{12}$ of any force acting between the members of the pair. The force between the members of the pair is F, taken positive when it tends to increase the distance of separation. Then by writing the equations for the members of the pair separately in the conventional form, it may be found, if the members of the pair are initially at rest with respect to each other, that the motion of the pair satisfies an equation which formally is exactly like the conventional equation for a single particle. The equation is:

$$F = M_{12} \frac{d^2 l_{12}}{dt^2}, \qquad \text{where} \qquad M_{12} = \frac{m_1 m_2}{m_1 + m_2}.$$

By working with three or more pairs and determining the corresponding joint masses, the masses of the individual members of the pairs, m_1, m_2, etc., may be found. Furthermore, the pair kinetic energy and the conservation law are exactly like the well-known expressions for individual particles. The pair kinetic energy $(\text{K.E.})_{12}$ is $\frac{1}{2}M_{12}(dl_{12}/dt)^2$, and the increment of this is equal to the work done by the pair force $F\,\Delta l_{12}$, which in turn is equal to the decrease of the potential energy $(\text{P.E.})_{12}$, which has the value $-Fl_{12}$.

It is especially to be noted that all the quantities which enter into the formulation of the pair behavior are directly determinable in terms of observations on the pair only and in no way involve the cooperation of an external frame of reference. Since we have seen that the conventional equations referred to a conventional frame of reference can be built up from pair behavior, we conclude that the introduction of a frame of reference is a matter of convenience only and not a matter of necessity.

The frame of reference to which we have been led by our considerations above is a "Galilean" frame. The operations by which we set up the frame contained no implication of uniqueness, and in fact an indefinitely great number of such frames is possible. The directions of the axes of any Galilean frame remain invariable in any other Galilean frame. That is, the Galilean frames do not rotate with respect to each other, and the origins of the different Galilean frames move without relative accelerations.

The Mach Principle

If two force-free particles in a Galilean system move so as to keep the direction of the line joining them constant over a finite time interval, this direction will continue to remain fixed. That is, if we look from one member of the pair toward the other, its direction in the coordinate frame will continue unchanged. This holds no matter how far distant the other member of the pair is. We can and do pass to the limit by letting the second body recede until it has reached the stars. Provided that the star is so far away that it exerts no appreciable effect on the body here (and certainly any direct gravitational action of a distant star on local bodies is hopelessly beyond present experimental detection), and provided that the body here has been moving directly toward or away from the star — a condition that is ensured by the requirement that the star exhibit no "proper motion" — then we may expect a local body to persist in moving in the direction toward the fixed star which it has once taken. Concealed here is all the physics involved in the invariable direction of the gyrocompass with respect to the stars, a phenomenon which has excited the wonder of generations of physicists and which has been loosely formulated in the "Mach Principle," according to which the inertial properties of local bodies are determined in some way by all the other masses in the universe, particularly the distant masses. It is, however, not necessary to assume any esoteric influence of the stars on terrestrial phenomena. The stars appear to have a connection because they are so far away that they can have no connection. (The matter is discussed further in my note "The Significance of the Mach Principle," *Amer. Jour. of Phys.*, **29**, 32–36 [1961].)

Nevertheless, I think there is something here which still may plausibly excite the wonder of the physicist. Suppose that in our chosen frame of reference there are two massive particles we have by some unspecified means brought to rest in a force-free condition. We now exert a force between the two members of the pair, preferably a repelling force, as by a negative thread or compressed spring. The bodies start moving apart. After a suitable relative velocity has been built up, we remove the force. *The bodies continue to separate for all future time at the same relative velocity.* This is the fundamental fact of mechanics which long acquaintance with Newton's laws of motion has made so familiar to us that we have ceased to wonder. Yet here is a veritable miracle. For in preserving a constant relative velocity, the bodies preserve a constant relation to each other. But how can such a relation be maintained? By hypothesis, the bodies are force-free and therefore without mutual interaction. Furthermore, they are, by hypothesis, moving through empty space where there is nothing which conceivably might mediate a distant connection. Yet somehow concealed in these particles there is a memory of their past connection and a prediction for the future of their coming relation. How can *particles* do so much, and what is a conceivable mechanism by which they might do it? One can understand the frame of mind of the Greek philosophers who had to have the continuous action of a force to maintain a body in uniform motion.

On the other hand, it is possible to find a point of view from which the wonder has partially evaporated. For after the force between the two bodies has been removed, the distance between them continues to *exist*, and this must be *some* sort of function of the time. The relative velocity also continues to exist and must also behave in *some* way as time goes on. What more natural or simpler way than for the velocity to continue to have the same value it did at the moment of cessation of the force? The principle of sufficient reason makes very plausible what actually happens. The wonder that remains is the wonder that there should be any regularity or lawfulness at all under these circumstances. It is obvious that the relative velocity of the receding bodies has to be *some* function of the time, but it is not obvious that whenever we repeat the experiment it always has to be the same function.

Galilean Frames

In actual physical embodiment, a Galilean frame is a rigid physical scaffolding of some sort to which a coordinate system can be attached, or some sort of base on which a coordinate scaffolding may be erected. The scaffolding need not be in all respects a material scaffolding. Given a sufficiently extensive base, the equivalent of a material scaffolding may be erected optically, and the coordinates of distant points determined by measuring the time required by light signals to reach them from the base. Any actual material structure which is a candidate as a base for a Galilean framework contains internal evidence as to whether it is suitable. The members of a Galilean frame are free from internal stresses. Whether there are internal stresses may be determined by cutting the suspected member and observing whether the cut members retain their relative positions. A less drastic method, applicable to some cases, is to test for internal stress by optical double refraction.

We have said nothing about the mass associated with a Galilean frame. There is no specific requirement here, except for the implication that associated with the particle there had to be *some* mass which served as the origin of the primitive set of three fixed directions. But if we expect to use the framework as an anchorage for the arbitrary forces we shall want to apply to the various particles in order to induce in them any desired state of motion, then we shall obviously have to make the frame massive as well as rigid. "Massive" means much heavier than any of the particles we expect to put into interaction with it. In practice, of course, this is what we have actually done. The frame we most naturally use is that of our laboratories, which are securely fastened to the earth, which thereby becomes the base of our coordinate system, a base obviously much more massive than any of the objects we ordinarily want to handle. The earth, however, does not provide a truly Galilean frame, and if we want to base a Galilean frame on the earth, we shall have to apply small corrections to measurements made in the frame of the earth. These corrections can all be obtained by observations completely self-contained on the terrestrial scene, such as the rotation of a Foucault pendulum or the deviation of a falling body from a plumb line. There is no need to bring in any reference to the stars, although it is often convenient to do so.

Two Frames of Reference

With this examination of the physics of a single frame of reference we pass to a consideration of two frames, one of these the "stationary" frame and the other the "moving" frame. By designating the frames in this way nothing more is necessarily implied than that the second frame is in motion with respect to the first (and vice versa). We have already seen that the operations by which the first frame is selected, which are such as to ensure only that it is "Galilean," are by no means unique, and that the one frame we arbitrarily selected and designated as "stationary" was chosen from a number of frames all moving relatively to each other.

The mathematics of the relations between the coordinates in the stationary and moving frames have already been displayed in the equations on page 7. In this chapter, we shall deal at once with the full three spatial dimensions, using the full set of equations, including $y = y'$ and $z = z'$.

As always, the relative motion of the two frames is along the x-axis.

Constitution of the Frames

The frame of reference implied in these equations must be a physical frame of some sort, since we expect to apply the equations to physical situations. In the stationary system, the frame may actually be constituted in a great variety of ways. For simplicity, we shall think of it as a three-dimensional scaffolding of rigid meter sticks, with clocks for determining the time of local events situated at the nodal points. We have already discussed the method by which these clocks are to be set. We have seen that we could dispense with the array of clocks and get along with a single clock at the origin. But this would involve correcting the readings of the single clock by a process which physically is completely equivalent to the process by which we set the array of clocks. Since we thus gain nothing essential by the use of a single clock, we shall adopt the conceptually simpler scheme of setting up an array of clocks at the nodal points.

Up to now we have not inquired very carefully as to the nature of the meter sticks and the clocks with which we construct the stationary frame, but have dismissed the question with the easy specification that the meter stick be "rigid" and that the clocks satisfy our intuitive demand as some sort of mechanism, the rate of which is independent of its location and the setting of which can be changed at pleasure. We shall return presently to a more detailed examination of the nature of meter sticks and clocks.

We now ask what sort of a thing physically is the moving frame. The answer is that it is the same sort of thing as the stationary frame, except that it is in motion. But what exactly does this mean? How can two frames be the "same" when one is moving and the other not? What shall we do to ensure that the two frames are the "same"? An obvious way to do this is to construct two frames initially side by side, which are identical by every possible physical test, and then to set one into uniform motion with respect to the other. Another way is to take the blueprints according to which the frame was constructed in the stationary system and, with them in our hands, transport ourselves to the moving system and there erect, *ab novo*, according to the same specifications, a new frame. Logically, perhaps the second course is to be preferred, and some writers choose it, but one can see that in actual execution embarrassing questions are going to arise. For example, how can one be sure that one can find in the moving system the *identical* material out of which to construct its meter sticks and clocks? Certainly when one applies the results of relativity theory to some actual physical situation, one does not use a moving frame constructed in any such esoteric fashion. Consider, for example, the Michelson-Morley experiment in that form in which the position of the central interference fringe is followed through a twelve-hour interval. During this interval, the velocity of the apparatus, whatever it may be "absolutely," has changed by a known amount because of the earth's rotation. At the initial observation, the two arms of the apparatus constituted a frame of reference. In the second position, twelve hours later, the same apparatus constitutes the moving frame.

In general, it will be found that the applications of relativity theory demand that the meter sticks and clocks of the moving system be the same physically as those in the stationary system, in the sense that they may be obtained by actual transport from the stationary system. The physical content of the transformation equations is contained in the behavior of transported meter sticks and clocks. The coordinates in the moving system are the coordinates supplied by such transported meter sticks and clocks.

Questions at once present themselves: How shall the transport be effected? Are its details irrelevant? The equations themselves contain no answer to questions of this sort, but such questions have to be answered by the "text" accompanying the equations. This text is seldom formulated explicitly and is not part of the recognized corpus of the theory. The reader usually has to construct the text for himself, by observing the way the theory is applied in practice. There is no way by which consensus can be automatically attained here, and the locus of most of the controversies which still plague relativity theory is just here, in the text.

Whatever the text, the equations by their mathematical form make certain commitments about topics with physical content — for example, the relativity of simultaneity, the contraction of a moving meter stick, and the retardation of a moving clock — topics we shall now consider.

Relativity of Simultaneity

Consider, in the first place, what the equations have to say about the relativity of simultaneity. It was mainly because of its pronouncements about simultaneity that relativity encountered obstinate resistance and misunderstanding when it was first formulated. The issue arises only in connection with events that occur at different places. If the events occur at the same place, then relativity theory treats simultaneity as a primitive concept, such that the judgment of simultaneity can always be made intuitively and uniquely. If the events occur at different places, then, *by definition*, they are simultaneous if the readings on the two clocks at the two places where they occur are the same. We have already seen in considering a single frame of reference that two clocks in two different places imply some method of speading time over space. To the extent that this method of spreading time may contain an arbitrary element, it appears that even in a single frame distant simultaneity must be relative, not absolute — relative to the method by which time is spread. This, however, was not the connotation when we spoke of the relativity of simultaneity above. The "relativity" there implied was a relativity to the coordinate frame, not a relativity with respect to the method of spreading time. The old intuitive notion, back of Newton's mechanics and all thinking in mechanics up to the time of Einstein, was that simultaneity of distant events was an "absolute" in the sense that it was independent of the coordinate frame. That is, the old notion was that two events distant from each other which were simultaneous in one frame of reference were also simultaneous in another frame moving with uniform velocity with respect to the first.

The equations at once provide an answer as to whether distant simultaneity is absolute or relative in this sense. Consider two distant events which in the stationary system are simultaneous. The first event shall have the coordinates $x_1 = 0$ and $t_1 = t_0$, and the distant event shall have the coordinates $x_2 = x_0$ and $t_2 = t_0$. Substituting in the equations, we find for the time of the first event in the moving system

$$t_1' = \frac{t_0}{\sqrt{1 - \frac{v^2}{c^2}}},$$

and for the time of the second event

$$t_2' = \frac{\left(t_0 - \frac{vx_0}{c^2}\right)}{\sqrt{1 - \frac{v^2}{c^2}}}.$$

These two times are obviously not the same, and the two events are not simultaneous in the moving system.

In general, therefore, two distant events which are simultaneous in the stationary system are, by definition, not simultaneous in the moving system (or, in other words, whether two events occurring at a distance from each other are simultaneous depends on the coordinate system in which they are described). Such a conclusion has proved highly unacceptable to the intuition nurtured in classical mechanics. The conclusion is so unacceptable that in the early days many reacted by refusing to accept the transformation equations, regarding acceptance of them as too high a price to pay. In fact, one *can* refuse to accept the equations, and by fiat retain the absoluteness of distant simultaneity, by simply replacing the second equation by $t' = t$. What is the objection to this, in view of the fact that we have already recognized that the way in which we spread time over space contains an arbitrary element.

There are two sorts of objection to spreading time over space in such a way as to yield this sort of absolute simultaneity. In the first place, the transformation equations so obtained would not be applicable to a number of other situations, as in electrodynamics, without introducing serious complications. In the second place, such a method of spreading time violates a cardinal principle, not only of relativity but also of any acceptable physics. For the method of spreading time in the second system is not the "same" as that used in the first. In the first system we spread time by using light signals in an easily describable way, whereas in the second system we use a hybrid method, involving first a spreading in the first system by light signals and then a transfer to the second system by a mathematical operation. In these operations the first system is thus singled out for special treatment.

The cardinal principle referred to above is that in our physical theorizing or analysis no individual object or system should be arbitrarily singled out for special treatment unless the system possesses some physical characteristic which sets it apart. Here we have singled out for special treatment the first system. This is a Galilean system, it is true, but it has nothing physical to distinguish it from the infinite number of other Galilean systems. The system we chose as the first system was the same only by luck; the second system might as well have been the first.

Consistency with this point of view demands that our sought-for "absolute" simultaneity be defined in terms of some method of spreading time that shall be the same for all systems in uniform motion with respect to each other. Relativity theory makes the simple statement that there is no such method. One might perhaps be inclined to question this categorical statement by examining in detail other possible methods of spreading time. We have seen that there are at least two other such methods. One of these involves the use of a sweeping searchlight with superlight velocity, and the other the use of clock transport with self-measured velocities and extrapolation to zero velocity of transport. Suppose that we spread time by moving the sweeping searchlight. What is involved when we apply the method to the moving system? We at once observe that we cannot use the *same* searchlight, but must use a second, which is stationary in the moving system in the same way that our original searchlight was stationary in the stationary system. We can see at once that this will so complicate matters that we lose interest in carrying out the computational details, and allow the matter to drop, in the conviction that if we did carry out the complicated details we would get the same result we did with the conventional method with light signals. Incidentally, it may be noted that concealed here is the reason why an infinite velocity of sweep in the stationary system is not infinite in the moving system.

If we try to spread time by the method of transported clocks, self-measured velocities, and extrapolation to zero velocity, we encounter a similar defeating asymmetry in the description of the process in the two frames of reference. In the stationary system, we transport clocks at different velocities all to the same place and make our extrapolation. In the moving system, the same physical procedure is going to land the clocks in different places, and the extrapolation is complicated, to say the least. If relativity is right, as we believe it is, then we must get the same result as before if the details of the transportation process are so arranged as to land the clocks transported with different velocities in the moving system in the same place in that system.

It thus appears that, as far as we can tell, the concept of the relativity of the simultaneity of distant events is an internally entirely self-consistent concept. However, I think that some persons will still not be able to suppress all yearning for the classical absolute simultaneity. I suspect that the explanation of much of this yearning is psychological and can be traced to the fact that we directly perceive local simultaneity as unanalyzable. In this sense, local simultaneity is an absolute. Psychologically, we strive to extend the range of applicability of the concept indefinitely. The extent to which we find this intuitive concept applicable to everyday happenings is as far as we can see on the terrestrial scene. Our perceptions are not sensitive enough to give us any direct inkling of the existence of propagation phenomena connected with light, nor do we at once appreciate that there are any logical problems in spreading time to a distance. It is our acceptance of the psychological present that causes our trouble.

It is of interest to inquire how distant simultaneity would appear if we allow ourselves the latitude in spreading time to a distance contemplated in Reichenbach's formula with a value of ϵ not equal to $\frac{1}{2}$. We can answer this by a simple modification of the analysis by which we deduced the Lorentz transformation on page 8. As before, we suppose the coordinates in S and S' are connected linearly, such that the origins coincide at the origin of time, and such that the systems are moving with velocity v with respect to each other. These conditions give for the transformation:

$$x' = a_1(x - vt),$$

$$t' = a_2 x + a_1 t,$$

where a_1 and a_2 are still to be determined. Now impose the condition that it is the average go-come velocity which is to be invariant and equal to c, not the one-way velocity, as in the canonical deduction. The go-come velocity involves Reichenbach's ϵ. A simple analysis, the details of which I do not give, yields the following condition:

$$\frac{a_2}{a_1} = -\frac{2\epsilon}{c} + \left(\frac{2}{c} - \frac{2\epsilon}{c}\right)\left(1 - \frac{2\epsilon v}{c}\right).$$

This condition contains v explicitly, unlike the former analysis. Now it is physically obvious that when v approaches zero, a_2 must equal zero in the limit and a_1 must equal 1. The equation shows that this limiting condition is not possible unless $\epsilon = \frac{1}{2}$, which is Einstein's value, corresponding to "isotropic" propagation.

It was a presupposition of the analysis just summarized that ϵ is the same in S and S'. Another way of stating the result is that if we have spread time in S so that ϵ is not equal to $\frac{1}{2}$, and if we are going to maintain the invariance of the go-come velocity of light in S and S', ϵ cannot be the same in S and S'. This ties in with another property of the transformation. Let us assume that $\epsilon \neq \frac{1}{2}$ in S, and call the velocity along the positive axis α and along the negative direction β. If the go-come velocity is c, we have

$$\alpha = \frac{c}{2\epsilon},$$

$$\beta = \frac{c}{2(1 - \epsilon)}.$$

If now ϵ and c are invariant in S and S', both α and β must also be invariant. That is, there are *two* invariant velocities on passing from S to S'. Now it can be shown in a few lines that there cannot be a linear transformation connecting S with S' which leaves *two* velocities invariant unless one of these velocities is the negative of the other. That is, only a velocity c in the positive direction and a velocity c in the negative direction can be simultaneously maintained invariant. But an α or β equal to c demands an ϵ equal to $\frac{1}{2}$, contrary to hypothesis. Furthermore, an ϵ not equal to $\frac{1}{2}$ and different in S and S' violates the first postulate of relativity, because S could be distinguished from S' on the basis of the numerical value of ϵ.

How now do these considerations react on the "conventionality" in the usual specification of distant simultaneity, or on Einstein's contention that he was "defining" distant simultaneity? In the narrower context of a single frame of reference in which time has to be spread to a distance by *some* method and it is recognized that this method may be arbitrary, adoption of the value $\frac{1}{2}$ for ϵ must be recognized to be a convention, and Einstein was justified in speaking of his distant simultaneity as embodying a "definition." But in the broader context of more than one frame of reference, the conventional element disappears. Einstein had no choice in the matter if he was to maintain the first postulate. There is, furthermore, a hard core of physical "reality" in the first postulate which compelled his choice. This hard physical core is the possibility of setting distant clocks in relatively moving systems consistently by different methods. We have followed through the details of three such methods: signaling with light, transporting the clocks, and the sweeping searchlight. That is, we have here the possibility of getting to the same terminus by more than one path. This possibility would seem to be pretty nearly synonymous with what the physicist understands by "reality."

Einstein's choice was also limited if he wanted to use conventional mathematics. This mathematics is a mathematics of extended time and, to the extent that any spreading of time over space is arbitrary, contains a "man-made" element. But we can hardly avoid spreading time over space or adopting a corresponding mathematics — we would find it hard to ignore the existence of the clock we can see on the distant church tower.

Reichenbach, apparently, did not consider the implications of a "conventionality" of distant simultaneity and a value for ϵ different from $\frac{1}{2}$ in the context of the first postulate — that is, in two relatively moving systems.

Contraction of Length

We next consider the contraction of a moving meter stick. This is explicitly contained in condition (5) on page 8 which is automatically fulfilled by the equations in virtue of their form. In the moving system, $x' = 1$ denotes the head of a meter stick moving with the system, the tail end of which is at $x' = 0$. The equations show that the head end of this meter stick is at the point

$$x = \sqrt{1 - \frac{v^2}{c^2}}$$

at the same instant, $t = 0$, at which its tail is at the point $x = 0$. These equations are the mathematical expression of the fact usually expressed by saying that the moving meter stick is shortened in the ratio

$$\left(\sqrt{1 - \frac{v^2}{c^2}}\right):1.$$

These equations contain the *definition* of the length of a moving object. By definition, the length, in the stationary system, of an object moving with respect to that system is the distance in the stationary system between the two points which the ends of the moving object *simultaneously* occupy. "Simultaneously" means simultaneously in the stationary system. This has to be specified because we have seen that simultaneity in the stationary system is not simultaneity in the moving system.

Some sort of *definition* of the length of the moving object is necessary, because there is no uniquely defined parameter or operation which passes naturally at the limit, at small velocities, into the well-defined length of a stationary object. In fact, there is another definition and corresponding operation of measurement a normal physicist would intuitively select as embodying what he would most naturally mean by the length of a moving object. If a physicist or engineer wanted to find whether the length of a streetcar is changed when it is set in motion, he would board the car with meter stick in hand and measure its length in the customary way, first when the car is standing and then when it is moving. If he did this, his result would not agree with that given by the other definition above. The physicist would, in fact, find no change of length at all by this procedure. Why then should the relativist go out of his way to adopt a definition which is more complex than one he might have adopted and which would not have offended physical intuition? The answer is that the relativist, being chiefly concerned with the *equations* of transformation, has been more interested in securing mathematical simplicity than he has been in securing simplicity of physical manipulation or description.

The definition the relativist has adopted for moving length, and which the physicist now accepts with complete good grace, is thus a definition which involves operations with clocks. Since there have to be two clocks, one at each end of the moving object, the result depends on the way in which we have chosen to spread time over space. We have adopted a method which involves the use of light signals. Does this mean that the "length" of a moving object does not have any physical "reality" by itself, but in some way is involved with the properties of light? This would be an unwelcome conclusion and would run counter to the commonsense belief we draw from everyday observation that the geometrical properties of objects are independent of their optical properties. In fact, this conclusion was repugnant to the intuition of Einstein himself, as well as of other writers, and Einstein has described a procedure for giving an "absolute" significance to the length of a moving meter stick, independent of any involvement with time or clocks. This procedure is to employ *two* meter sticks, one moving from right to left and the other moving from left to right with the same velocity. As the two sticks pass each other, the two left-hand ends will be in coincidence at some instant, and the two right-hand ends will also be in coincidence, whether later or earlier we do not have to determine. We mark in the stationary system the *point* at which the two left-hand ends are in coincidence and also the point at which the two right-hand ends coincide. This can be done without question and uniquely, since local coincidence is an "absolute," the same in all reference systems, and in no way involves the use of a clock. We now measure at our leisure in the stationary system the distance between the two points we have thus marked, and this distance gives, by definition, the length of the moving stick. By symmetry the length of the two moving sticks is the same. On working through the details, it will be found that the contraction of length obtained in this way is indeed the same as before.

Although this procedure apparently avoids reference to any clock, I think it is only apparent. For how can the equality of the velocity of the two sticks, the one to the right and the other to the left, be established except with the aid of clocks? As long as the definition of a moving length is a two-clock definition, I do not see how *some* involvement can be avoided with the way we have chosen to spread time, and the unwelcome conclusion referred to above apparently has to be accepted. The length of a moving object is not something that exists in its own right, in and by itself. It is therefore not appropriate to ask whether the contraction of a moving meter stick is "real" or only "apparent." Neither is it permissible to ask whether the "cause" of its contraction is its being set in motion. How shall we say whether it was the motion or the method of spreading time that was the "cause" of the shortening? The situation here is more complex than the situations to which the notion of causality can be unambiguously applied.

It might appear to an impatient intuition that we have been making altogether too much fuss and feathers about the moving meter stick. Could we not arrange matters so that we could *see* whether the stick is shortened or not? Could we not have the stick moving at a considerable distance from us at right angles to the line of vision in front of a screen on which we had previously marked out distances in the line of motion? We could then take an instantaneous photograph of the moving stick and the screen back of it and tell by subsequent inspection of the photograph whether the distance between the ends of the stick was one meter or not. This procedure would doubtless yield a result, but the result would not be what we want. For the same considerations are involved that we saw were involved when we tried to get a one-way velocity for a beam of light, also moving at right angles to the general line of vision and at a great distance. It is not possible by any such procedure to avoid the insidious entry of the way in which we spread time over space.

We have seen that there is an involvement of the length of moving objects with clocks which appears to be unavoidable. This does not mean that there is a necessary involvement with the properties of light, although this is the implication in some of the conventional discussions. It is true that we have defined the setting of distant clocks in terms of light signals, but we need not have so defined them. We could, for example, have used the method of clock transport. In any event, there has to be *some* formal involvement of moving length with other things — what the other thing is, is not uniquely determined. Because any acceptable method of spreading time must yield the same result as spreading time by light signals, we may perhaps want to say that any involvement with other things is basically an involvement with the properties of light. This is to a large extent a verbal matter and within our choice. The essential physical fact here is that other methods of speading time give the same result as light, and that any one of these methods may be made the basis for the transformation equations.

Nature of a Meter Stick

Let us now turn to the questions: What is a meter stick? How shall it be set into motion when we pass from the stationary to the moving frame? These questions are almost never seriously considered, but we are usually content with the conviction that the ordinary "rigid" bodies of everyday life function to a sufficiently good approximation as meter sticks when we use them in such an apparatus as a Michelson-Morley interferometer. The basis for this conviction is, of course, that the equations of relativity are consistent with the Michelson-Morley experiment. As a practical matter for the working physicist this may be satisfactory, but as a matter of cold logic it is obviously highly unsatisfactory. That the matter is not simple and that there are very real issues here is made obvious by Dewan and Berans' recent paper "Note on Stress Effects Due to the Relativistic Contraction" (*Amer. Jour. Phys.*, **27**, 517–18 [1959]). They call attention to the fact that if two equal masses separated by unit distance along the *x*-axis and originally at rest are set into motion by the simultaneous action of two forces always equal to each other, their accelerations and therefore their velocities will at every moment be the same. In consequence, the distance apart of the masses will remain unaltered. This, of course, is a consequence of the fundamental assumption that absolute position is without effect on physical phenomena — one point on the axis is as good as any other for the origin.

The application of this observation of Dewan and Beran to the problem of the meter stick is immediate. There would seem to be no reason why the two masses, as long as they remain force-free, should not themselves serve as a meter stick, marking out the position of the ends of the meter. We could, if we like, subdivide our meter stick into centimeters by stationing independent masses at centimeter intervals, or we could proceed to the limit with a continuous distribution of mass. Our considerations apply only so long as we set the continuous distribution of mass into motion by the simultaneous application of local forces proportional to the local masses, and so long as our "meter stick" has been set into motion with no change of length. This is a most disconcerting situation, for we have here an explicit receipt for the construction of a meter stick and detailed directions for setting it into motion, but there is no change of length as relativity theory demands. These specifications for the construction of the stick and for setting it into motion are about as simple as can well be imagined. Furthermore, these specifications give self-consistent results when we apply them in a single frame of reference. For by a suitable variation of the transporting force with time, at first an accelerating force and afterward a retarding force, the masses which mark the ends of the meter may be transported to any distance in the stationary frame and brought to rest again there, where they again mark out a meter. That is, this method secures the transportation of meter sticks to a distance with no change of length. The possibility of this is a fundamental assumption about the stationary system.

The actual carrying out of the transportation of our meter stick is not altogether simple. We are required to apply equal forces *simultaneously* to the two masses. This cannot be done on the spur of the moment, but requires previous preparation. We have to make previous arrangements with an assistant stationed at the second mass, or we have to rig up a suitable automatic gadget, so that when the clock at the second mass reaches a preassigned figure, a force is applied equal to the force we apply when our clock reaches the same figure. But, in principle, there would seem to be no reason why our transportation procedure should not involve previous preparation, or why our proposed meter stick should not be acceptable.

Relativity theory works in practice, and its meter sticks are doubtless shortened when set into motion. This means that the details of the way in which actual forces act when actual meter sticks are set in motion are different from the details of our specification above. Furthermore, the difference is definite and universal, for most actual materials serve to a high degree of approximation as possible raw material for the construction of meter sticks. Ordinarily, when we move our meter stick, we are not concerned as to which particular point of the stick it is to which we apply the force that moves it. Neither are we concerned with the obvious fact that every act of transport, in practice, involves the generation of elastic waves in the material of the stick, which are later damped out by some mechanism to which we pay no attention. The fact that all our meter sticks, regardless of their material, are insensitive to these things indicates that there must be something deep-seated in the structure of all matter which secures its conformity to the principle of relativity, something too subtle to be readily formulated. Doubtless there is a tie-in here with the fact that the most important applications of relativity theory are in the realm of microscopic phenomena.

It would seem that no reduction of relativity theory to an axiomatic postulational basis can be complete without the addition of the postulate that actual meter sticks act in accordance with the Lorentz equations.

Retardation of Clocks

We pass now from a consideration of meter sticks to a consideration of clocks. The retardation of a moving clock is a consequence of the form of the transformation equations and is contained in condition (6) on page 8. This equation has to do with the clock at the origin of the moving system, transported with the system. According to the equations, when this clock was at the origin of the stationary system it indicated the time 0, as did also the clock in the stationary system. When the hands of the clock in the moving system have moved to $t' = 1$, the time indicated on the corresponding clock in the stationary system is

$$t = \frac{1}{\sqrt{1 - \frac{v^2}{c^2}}}.$$

This is greater than 1, which means that the hands of the moving clock are not moving "as fast" as the hands of the stationary clock, or that the moving clock is retarded with respect to the stationary clock. This applies not only to the particular instrument we have chosen to call a "clock" in the moving system, but also to every other gadget that functions as a clock.

What sort of a concept is this "rate of a moving clock"? We would expect that it is a two-clock concept because obviously two clocks are involved — the moving clock and the stationary clock with which it is compared. But it is worse than this, and is actually a three-clock concept — one moving clock and two clocks in the stationary system, one at the beginning and the other at the end of the run of the moving clock. The two clocks in the stationary system again involve the way in which we spread time over space. It follows that the meaning of "rate" of a moving clock is a matter of definition, and that the element of definition is even more important here than it was with regard to distant simultaneity. For here we have to define a procedure with *three* clocks, whereas formerly we were concerned with only two. Hence the question as to whether the moving clock is *really* or only *apparently* slowed is as illegitimate as the question as to whether a moving meter stick is *really* shortened or not. Similarly, is it an illegitimate question to ask whether the *cause* of the retardation of the moving clock is that it is in motion, or whether it is the acceleration that necessarily accompanied its being set into motion? The concept of causality with all its implications does not apply to this situation.

A clock is a more complicated physical apparatus than a meter stick because a clock has two controllable parameters instead of one. The "clocks" with which we are familiar in daily life have provisions for adjusting either the rate or the setting. A clock is essentially an apparatus for measuring the time interval from some zero. Changing the "setting" amounts to changing the origin from which the interval is reckoned. We ordinarily do this by changing discontinuously the positions of the hands of the clock. We could, however, adopt a different procedure. Instead of actually changing the position of the hands, we could apply mentally a correction to the observed reading of the clock, thereby effectively changing its setting. We often do precisely this as a temporary expedient, but because of greater convenience we usually end by changing physically the positions of the pointers of our timepieces.

It is a matter of observation and experience that the setting of the watch in one's pocket does not change if one suddenly set it into motion. In fact, this observation about the setting of our timepieces is so inextricably interwoven with our entire mechanical experience that we are almost tempted to argue that the setting *could* not change. For if the setting were to change abruptly, this would mean a sudden change in the *position* of the hands, which would involve infinite velocities and infinite forces if the hands had a finite mass, as they do.

This all has relevance when we consider the clocks in the moving system. For we have seen that the moving frame has to be the same physically as the stationary frame, and is generated by setting a stationary frame into motion. It follows that when we generate the moving frame from the stationary frame there is no alteration in the settings of the clocks. There is, however, an alteration in the rates. What we mean by this "rate" is a matter of definition. We have in effect already implicitly made our definition, and in accordance with it the numerical measure of the alteration of rate is that already given on page 98, namely

$$\frac{1}{\sqrt{1-\frac{v^2}{c^2}}}.$$

This change of rate takes place instantaneously, as soon as the clock is set into motion. No finite displacements or infinite velocities and infinite forces are involved in such an instantaneous change of rate.

After allowing our clock to move with the moving frame for a while, we may remove it from the moving frame and establish it again in the stationary system, at a distance from its original position. The clock thus transported to a distance will find waiting for it a clock which has been set according to the procedure dealing with a single stationary system already described in Chapter Two. The settings of the two clocks will disagree. It is the thesis of relativity that the amount of disagreement of the settings can be entirely accounted for in terms of the change of rate of the transported clock while it was undergoing transportation. Since the moving clock runs "slow," we shall find, if our definition of the rate of a moving clock is a good one and if our thesis is correct, that the hands of the transported clock lag behind the hands of the clock which it finds waiting for it, so that the setting of the transported clock will have to be advanced to agree with that of the local clock. The self-consistency of the whole fabric of relativity theory leads to the belief that this is what happens in fact.

Now let us reverse the process and transport the clock we have just carried to a distance back to the origin again.

We do this by incorporating it for a suitable length of time in a second moving system, this one moving with exactly the same velocity with respect to the stationary system as the first but in the opposite direction. The second transport is symmetrical in every way with respect to the first transport. According to our formulas, the retardation involves the *square* of the velocity of transport and is therefore the same numerically on the return trip as on the outward journey. It follows that the setting of the twice transported clock will again have to be advanced to agree with the clock at the origin. If the transported clock is taken through the entire round trip with no change of setting at the distant station, it will return to the origin lagging behind the clock which remained stationary at the origin by the sum of the two amounts.

Space Traveler

Concealed here is the so-called paradox of the "space traveler" which has aroused so much recent controversy. If we may assume that a biological system functions as a "clock" — and there would seem to be no good reason why it should not — then the conclusion is that a traveler who returns to his base after a round trip through space comes back younger than his identical twin who remained at home. Such a state of affairs has profoundly offended the physical intuition of some people (Professor Herbert Dingle conspicuously), who have seen here a violation of the cardinal principle of relativity itself. One may indeed sense a paradox here if one subscribes to a rather vague intuitional formulation of the principle of relativity. For if all things are relative, and in particular if motion is relative, then one might equally well say that it was the first twin who had stayed at home and the second twin who had gone away and returned. The implication here has to be that the twins themselves could not tell which was which, and that either would use exactly the same language to describe his own experiences. But this is patently not the case. For the twin who went and returned has twice transferred from one system to another in relative motion with respect to it, and this transfer, despite the fact that there was no change in the setting of the watch in his pocket, had other physical manifestations — for example, two quite unforgettable jolts when the transfers were effected. The experiences of the two twins were by no means the same, and the difference of their experiences is uniquely correlated with the difference in their ages which they discover on reunion. There is here no violation of any formal principle of relativity.

There are some who would maintain that there can be no paradox here, for the reason that relativity theory is logically applicable only to systems in *uniform* relative motion and that it can have nothing to say about phenomena involving such accelerations as are obviously inextricably involved when a traveler goes away and returns. The answer is that although in its formal axiomatic setup relativity theory may not deal with accelerations, in application it cannot avoid doing so, because the meter sticks and the clocks of the stationary and moving systems have to be the same, and this involves transport and accelerations. Accelerations would seem to enter the arguments above with respect to the space traveler to no greater extent than they enter the analysis of the Michelson-Morley experiment, which all agree is covered by the special theory.

It is not necessary to pursue this topic further here. In

the published arguments, many detailed analyses have been given of various aspects of the situation in the endeavor to show a "causal" connection between the various factors which enter into the final difference of age, of which the "jolts" mentioned above constitute a special case. It is not necessary to analyze the merits of the various analyses, but the comment may be made that it is not probable that all the implications of a causal point of view can be carried through with regard to all the details of this complicated situation. It makes no sense to ask whether the returned twin is "really" younger than his stay-at-home brother.

An interesting question is whether the returned space traveler can give a completely causal account of the difference of aging which the stay-at-home ascribes to him in terms only of things which have "physical reality" for the traveler — that is, in terms of the magnitudes of the jolts which he receives and his self-measured time intervals. A hasty preliminary examination indicates that the answer is probably No — the various significant factors do not have a sufficient degree of linearity. It is to be noticed in this connection that the "jolt" received on starting the return journey is not, because of the addition law of velocities, twice the jolt on starting, but is less by the factor $1 + v^2/c^2$.

Nature of a Clock

In discussing meter sticks and their transport from the stationary to the moving system, we found it unsatisfactory that there is no formal definition of what constitutes a meter stick. We had to leave it with the rather vague remark that the approximately rigid bodies of ordinary experience function to a sufficient degree of approximation as meter sticks.

The situation is somewhat similar with regard to clocks. No precise definition of a clock is usually attempted, and we are usually satisfied with the vague conviction that such things as the vibrating atoms that were the source of light in the Michelson-Morley experiment do in practice function as "clocks." In fact, a personal conversation with Einstein has been reported to me in which he said that his theory stood or fell with the assumption that light-emitting atoms do in fact constitute clocks. It is, however, not necessary to leave matters in quite as vague a condition as this, for detailed specifications can be given for at least one clock.

Instead of attempting to specify our clock in mechanical terms, which would involve difficulties because the postulates of relativity theory do not immediately deal with mechanics, we can give our specifications in terms of the properties of light. Our clock shall be a nonrotating meter stick, with mirrors at the two ends, between which a pulse of light is continuously reflected back and forth. Time is measured in terms of the number of passages back and forth or in terms of the number of reflections at either mirror. This is, of course, a fantastically impossible sort of clock from a practical point of view, but that need be no formal objection. It is obvious that this "light clock" functions in the stationary system as a clock, for meter sticks can be transported with no change of length, clocks can be transported to new stationary stations with no change of rate, and the velocity of light is everywhere constant. It is also obvious, and superfluous to try to prove, that the same device (nonrotating meter stick with mirrors) also functions as a clock when transported to the moving frame, for the invariance of the velocity of light in moving systems has been built into the structure of the transformation equations.

Having before us this specific clock, to which the readings of every other "clock" must conform, it appears more plausible than ever that no device can be constructed for measuring time intervals which does not involve in some way with the measurement of space, for our specific clock explicitly contains a meter stick.

The transformation equations contain the answers to several other questions we have not yet considered in detail. For example, they contain the answer to the question "What is the self-measured velocity of light?" To answer this, we have to put $v = c$ in the Lorentz equations, and find how t' compares with t for some appropriately chosen x. The equations show at once that $t' = 0$ for every x and t. That is, the clock transported with the velocity of light gets to its distant station with no change of its reading. In other words, the self-measured velocity of light is infinite. In general, the self-measured velocity of a clock transported with velocity v is

$$\frac{v}{\sqrt{1 - \frac{v^2}{c^2}}}.$$

(This may be found at once from the equations by finding x/t' when $x' = 0$.) The self-measured velocity differs from the velocity measured in the stationary system by the same factor that the rate of a moving clock differs from the rate of a stationary clock. We, who stay at home, can therefore give a consistent account of the "aging" of the space traveler, after he has returned home, in terms of the effect of his motion on the rate of his clock, as described by us. It is to be noticed, however, that "age" is properly self-measured age — that is, "age" is properly a self-measured thing.

In cosmological speculations, one not infrequently hears talk of the "age" of a beam of light. For instance, it has sometimes been speculated that the cause of the apparent shift toward the red of light from distant parts of the universe is merely an effect of the degeneration with age of the light on its way to us. But if the self-measured velocity of light is infinite, its "age" is always zero, no matter how far distant the source. If one could ride along with a beam of light, there would be no light.

Addition of Velocities

Another topic covered by the equations in their conventional form is the composition of velocities. The mathematical relations have already been given and are contained in condition (7) on page 8. The formula there given is usually described as the "formula for the composition of velocities." The meaning is taken to be that a velocity u "added" to a velocity v does not give the classical $u + v$, but instead the smaller quantity

$$\frac{(u + v)}{1 + \frac{uv}{c^2}}.$$

Another way of saying it is that the velocity relative to the stationary system of an object moving with the velocity u in the moving system is as given by the formula. This formula has been the occasion of much unnecessary misunderstanding and paradox. The misunderstanding could largely be avoided by noticing that u and v are velocities in different coordinate systems and are therefore measured with different instruments, so that "relative" has in consequence experienced a subtle shift of meaning. In the moving system, u is the velocity and is measured by x'/t', whereas v is the velocity of the moving frame as measured in the stationary frame and is defined as x/t. Of course, if both velocities were measured in the same frame, the classical simple addition formula for relative velocities would continue to hold. In particular, in the stationary frame the velocity of a particle moving to the left with velocity $0.75c$ relative to a particle moving to the right with velocity $0.75c$ is $1.5c$, as it always has been and always will be, despite frequent statements that relative velocities higher than c do not occur. Such a "relative" velocity is obtained by calculation, not by direct physical measurement. If we stationed a measuring apparatus on the particle moving to the left with velocity $0.75c$ and with it measured the velocity of the particle moving to the right with velocity $0.75c$ we would obtain something less than c. In fact, we would obtain $0.96c$.

In a single system, the upper limit for the relative velocity of two beams of light or of two material objects is not c but $2c$.

The addition formula has to be used with care, in the context of its definition. The formula is of special service in showing the unique properties of the number c in the formulas. For if we put $u = c$, then the "sum" with v is identically c, no matter what v. That is, anything whatever, light or anything else, which in the moving frame has velocity c also has the velocity c in the stationary frame. Furthermore, anything moving with less than velocity c in the moving frame is also moving with less than velocity c in the stationary frame. In saying this we are not primarily saying something about light but are saying something about the way the number c enters the equations. It is true that because of our association of c with light we think of the equations as saying something essential about light, but this is not forced by the mathematics. Any fundamental role that light plays in relativity theory comes by way of the text, not by way of the equations themselves. But because of the text, the equations do purport to describe the behavior of light and in this way make contact with experiment.

General Structure of the Theory

Scope of the Theory

The equations as they stand are capable of handling all the phenomena of light propagation customarily included under geometrical optics. Some of the physical phenomena which thus come within the scope of the equations are: the aberration of light from a distant source, the Fresnel dragging coefficient, and the first and second order (transverse) Doppler effects. The discussion of these phenomena in the standard texts is straightforward enough, and we need not repeat it here.

As they stand, the equations do not mention many things that would be necessary if they were to be applicable to the complete range of physical phenomena — such things, for example, as mass or force or electrical charge. Hence, as they stand, the equations are not applicable to the ordinary mechanical phenomena of dynamics. Neither are they adapted to deal with electrodynamics. They have, however, a partial application in the form in which they stand. If, for example, we have some particular mass moving under some particular impressed force, then the Lorentz equations state that the coordinates of the path of the mass in the stationary frame will be related to the coordinates in the moving frame in the way specified by the equations. The Lorentz equations do not, however, enable us to say what the path is. However, relativity theory is ostensibly a complete theory and should provide means for calculating what the path is. There are means for doing this, and these means are part of the essential content of relativity theory. But these means are not explicitly contained in the equations, which have to be supplemented in some way. My position here is opposed to that of many writers, who see the Lorentz equations as containing the entire content of the theory. G. Builder, in the *Australian Journal of Physics*, **11**, 458 (1958), makes a particularly explicit statement of this position, which I quote in its entirety:

> To avoid any confusion in the ensuing discussion, it is necessary to state precisely what we will mean by the restricted theory of relativity. There is only one statement of this theory that can command universal assent, that is subject to experimental verification, and that is equally appropriate whether one ascribed the theory to Poincaré and Lorentz or to Einstein, namely "The restricted theory of relativity is the theory that the spatial and temporal coordinates of events, measured in any one inertial system, are related to the spatial and temporal coordinates of the same events, as measured in any other inertial system, by the Lorentz equations."

A position such as this can be given apparent justification by quotation from some of the expositions which consensus has accepted as authoritative. Thus, the following statement will be found on page 21 of the 1958 English translation of W. Pauli's *Theory of Relativity:*

> We showed in Part I that the two postulates of relativity and of the constancy of the velocity of light can be combined into the single requirement that all physical laws shall be invariant under the Lorentz transformation.

But a critical reading of this statement shows that it does not justify a position like that of Builder. For Pauli is saying only that the Lorentz equations can be derived from the two postulates; he is not saying that the relations can be reversed and the two postulates derived from the transformation equations. (It is to be remarked incidentally that the quotation in translation from Pauli as given above is not a very good translation, for it gives a wrong implication. A better translation would be: "We showed in Part I that the two postulates, namely the postulate of relativity and the postulate of the constancy of the velocity of light, can be combined. . . ." There are not *two* postulates of *relativity.*)

The mere observation that the equations as they stand do not offer means of coping with the problems of dynamics or of electrodynamics seems to me to constitute sufficient refutation of as extreme a position as that of Builder and others.

It would appear that the time has come when we should adopt a broader base than has sufficed hitherto and consider the whole structure of the theory. This means, in particular, that we should formulate the two postulates which are at the bottom of the theory, find what they mean, find in what way they enter the Lorentz equations, and find what their other consequences may be.

The Postulates of Relativity Theory

There are nearly as many formulations of the postulates as there are authors. All of these formulations involve more or less intuitive concepts, and the meaning has to be analyzed. All formulations present a first general postulate, usually called the "postulate of relativity," and a second particular postulate having to do with the properties of light. All formulations are concerned with two different coordinate systems or frames of reference, one moving with uniform velocity with respect to the other. The formulation of the first postulate given by Einstein (*Relativity*, New York: Holt, 1921, p. 15) is typical and is as follows: "Natural phenomena run their course with respect to K' according to exactly the same laws as with respect to K." The system K has been previously defined as a Galilean system, and K' is any other system moving with uniform velocity with respect to K and is also a Galilean system. This postulate is sometimes paraphrased as follows: "The laws of nature are the same in all Galilean frames of reference." Other formulations designed to cover the same universe of discourse may invoke the impossibility of performing any measurement which will detect "absolute" velocity. Such formulations are objectionable from a logical point of view, and they become understandable only in the historical perspective in which "absolute" motion means motion with respect to an "ether." All formulations require amplification, even Einstein's, for we have to know what a "general law" is, and what it is to run the "same" course in two coordinate frames.

The second postulate, being particular, is easier to formulate, and different authors do indeed essentially agree. The second postulate is that the velocity of light is independent of the velocity of its source. It would seem to be mostly a historical accident that two postulates were used instead of one. The first postulate leaves undefined what is to be understood by a "law of nature." This is logically unsatisfactory. In default of a general definition, the first postulate is meaningful only if what is to be understood by a "general law" is made specific by detailed enumeration. One, and perhaps the most important, of the "laws of nature" which would thus be specifically enumerated is that the velocity of light in empty space is the same in all Galilean reference systems. This general statement contains no reference to the nature of the source of light — whether it is in motion or has some other distinguishing characteristic — so that by implication the velocity of light is independent of the velocity of the source, and the second postulate is superfluous. Notice that the converse does not follow. The velocity of light might depend on the velocity of the source and we might still have a general law for the velocity of light, valid for all reference systems — namely, that the velocity is compounded of the velocity of the source and some characteristic velocity, in the manner of a projectile. This suggests what may historically be the explanation of the two postulates instead of the logically simpler one, because when the theory was first formulated the acceptability of Ritz's ballistic theory of light was under serious discussion.

In the usual handling of the second postulate there is a lack of logical precision which is probably not of great importance. The usual implication is that the velocity of light as measured in any frame is completely independent of the nature of the source, provided only that it emits "light." That is, the velocity is independent of the frequency, intensity, state of polarization, coherence, or motion. The last parameter, "motion," according to the usual implication may be anything, accelerated in any way or not. Logically this is probably demanding more than we need, and we could get along with the postulate that the velocity of light is independent of any *uniform* motion of the source. It could then be left open to experiment whether in fact there may be any effects due to acceleration of the source. As far as we know, there is no reason to expect the existence of such effects, except insofar as they may be connected with intense gravitational fields.

Events

We turn now to an examination of some of the implications of the fact that the principle of relativity is concerned with *two* reference systems. What is a "frame of reference" anyway? Observation of what we do with our frame of reference shows that it is primarily a coordinate system by means of which we associate characteristic numbers, the coordinates, with "events." These coordinates give the place and the time of the event. It is useful to know the coordinates of events, because these give a partial description which is applicable for many purposes. The description afforded by the coordinates is, however, by no means a complete description, for, given the coordinates only, the event could not be reconstructed.

What now is this "event" of which we give the coordinates? Examination of usage shows that the "event" is a concept of great generality, applying to many different sorts of physical situation. However, in all its usages it always has a temporal connotation and implies a "happening" of some sort. We are not likely to speak of a book passively resting on a table as an "event," although certain aspects of this situation may be referred to as events. Furthermore, and perhaps most importantly, an "event" exists in its own right, independent of the frame of reference or coordinate system in which it is described. It makes sense to talk about the "same" event in two frames of reference. The event is prior to the frame of reference and is independently describable and knowable, regardless of the particular frame which happens to have been chosen.

One of the simplest sorts of event is a coincidence. Coincidences may be of various kinds — two massive particles may collide, or an electron may produce a flash of light on the screen of a spintharoscope, or a moving object may be made momentarily visible by a flash of light. These are all coincidences, and they all have "absolute" significance, independent of the particular frame of reference in which they are described. There is no limit to the number of events which may be in coincidence — events are not mutually impenetrable like the atoms of Lucretius. The relation of coincidence is transitive; if A is in coincidence with B and B is in coincidence with C, then A is in coincidence with C.

The coordinate quadruple of the transformation equations in itself designates a special kind of event. The quadruple x, y, z, t designates the event of the hand of the clock situated at x, y, z in the stationary system reaching the point t on its dial. In the same way, the quadruple x', y', z', t' marks another event, the arrival of the hand of the clock at x', y', z' in the moving system at the mark t'. The transformation equations make a statement about the coincidences of these two classes of events. These equations have real physical content, for the two frames of reference are *physical* frames, constituted of a scaffolding of meter sticks with clocks at the mesh points. As these two frames move past each other, the mesh points of the one will be in a continuously changing succession of coincidences with the mesh points of the other. The transformation equations state that these coincidences will in fact be found to accord with the equations. Since now an event is an event, any other sort of event that coincides with the mesh point x, y, z, t of the stationary system will also be in coincidence with the corresponding mesh point x', y', z', t' of the moving system.

What then is the physical content of the theory of relativity if we try to formulate it by saying that the coordinates in the two frames of events of every sort are connected by the transformation equations? Are we really saying anything? Given the two frames, how could the coordinates in them help being related according to the equations? This does indeed seem to me to be a just criticism; we are not saying anything new when we formulate the theory in this way. The physical content of this formulation has already been exhausted in saying that the two coordinate frames are in fact related according to the equations. But obviously there are other things that need to be said here — we have already commented that the equations as they stand cannot cope with dynamics or electrodynamics.

To cope with these other matters, we have to make more explicit use of the first postulate. If we want to treat dynamics, we have to ask what the first postulate says about dynamics; and if we want to include electrodynamics, we have to ask what it says explicitly about electrodynamics. What now does the postulate say about these matters?

Logical Structure of the First Postulate

We must notice in the first place that the postulate does not say what at first glance it might appear to say. Consider the rough formulation of the postulate in the form "The laws of nature are the same in all Galilean frames of reference." A natural interpretation of the meaning would be that we are here stating a new observation about the properties of Galilean frames. The implication would seem to be that "laws of nature" are something that we know about from previous experience, and that the theory has given a new insight with regard to the relation of these laws to Galilean frames. This insight is that the laws of nature are the "same" in different Galilean frames, a statement that can be checked by formulating the laws for different frames of reference and observing whether they are the same. But this is not what the postulate is actually saying. What it is saying is something about a "law of nature"; it is saying that unless the "law" is the same in the two frames of reference it is not a "law." A better insight can be gained into this situation by considering an example in detail. In the analysis of the next few pages we consider only a single spatial dimension.

Conservation of Momentum

Consider a simple situation in mechanics, the collision of a mass $2m$ moving to the right with velocity w with a mass m moving to the left with the same velocity w. This problem can be solved by Newtonian mechanics. According to Newtonian mechanics, momentum is conserved in a collision, where momentum is defined as the product of velocity and a parameter characteristic of the individual body which is called the "mass." This mass is determinable by specific physical operations, which we assume are known. Any possible behavior of the bodies on collision is subject to the condition of conservation of momentum as thus defined. To determine completely the behavior of the above two masses after collision, one other condition is necessary. We shall assume that the collision is "elastic," which means that the kinetic energy is conserved in the collision — the kinetic energy of a single body being the product of its characteristic mass and one-half the square of its velocity.

The solution of our problem can be carried through in detail, and it will be found that after the collision the bodies both reverse the direction of their former motion, and that the mass $2m$ is now moving to the left with velocity $\frac{1}{3}w$ and the mass m is moving to the right with velocity $\frac{5}{3}w$.

But now this same collision can be described with the coordinates of the moving system, for the coordinates of the two systems are related by the transformation equations. The following state of affairs will be found, where the moving system is denoted by S', and v is the velocity of S' relative to S. Before the collision:

$$\text{velocity of } 2m \text{ in } S' = \frac{w - v}{1 - \frac{vw}{c^2}};$$

$$\text{velocity of } m \text{ in } S' = \frac{-w - v}{1 + \frac{vw}{c^2}}.$$

After the collision:

$$\text{velocity of } 2m \text{ in } S' = \frac{-\frac{w}{3} - v}{1 + \frac{vw}{3c^2}};$$

$$\text{velocity of } m \text{ in } S' = \frac{\dfrac{5w}{3} - v}{1 - \dfrac{5vw}{3c^2}}.$$

It is now the work of a moment to multiply each of these velocities by the corresponding mass, thus forming the various momenta, and to check whether in fact the total momentum in S' is the same before and after collision. It will be found that it is not. Does this mean that the first postulate has failed? By no means; it merely means that we did not understand what the first postulate was saying. What it was saying here is that it is not a "law of nature" that Newtonian momentum is conserved.

The first postulate thus appears to contain a concealed definition of "law of nature." By applying the criterion of the first postulate, we can determine whether an ostensible law is really a law or not. In general, the situation is as follows. We start with some concrete physical setup, such perhaps as a system of massive particles interconnected with a network of stretched elastic cords. This physical system develops in time, and the development in time is the chain of events in which we are interested. This chain of events can be described in either the stationary or the moving frame of reference — that is, in either S or S'. If we know the laws of mechanics, we can calculate the development of the system in S and thus find the final configuration in S, given the initial configuration. To this final configuration in S there corresponds a final configuration in S', determined by the transformation equations. But we could also start from the initial configuration in S', applying the laws of mechanics in S' — which in accordance with the first postulate we must assume we know — and calculate the development of the system in S', and thus find the final configuration. But we now have apparently said too much, for we have here two independent methods of arriving at the final configuration in S', and there would seem to be no a priori reason why the two configurations should agree. The first postulate states that it is not a "law of nature" unless the two methods of arriving at the final configuration do agree.

Without amplification, these considerations appear to have led to a completely sterile cul-de-sac. Who cares whether we used a "law of nature"? The initial and the final configurations of the physical system are simple "facts" over which we have no control — "absolute" in the sense that they are independent of any frame of reference, and describable by exhibition if in no other way. We are interested in acquiring control of this factual situation to the extent of being able to predict the way in which the system unrolls in time. We attempt to acquire this control through the formulation of some method which will allow us to make the pertinent calculations. The first postulate interjects a purely verbal factor into this situation by refusing to call this method a "law" unless it can be formulated for the frame of reference S' in the "same" way as for the frame S.

Nevertheless, I think every physicist will feel that there is something more than the purely verbal in the concept "law of nature" toward which we are groping. The most sweeping implication in "law of nature" would seem to be a degree of determinism in the way in which a chain of events develops in time. If we are in command of the law, we are able to formulate some algorithm by which we can correctly calculate the development in time. We have in the past been successful in finding many such algorithms and in calculating the development in time of many different sorts of physical situations. We are thus convinced that laws of nature "exist." The algorithm in terms of which any specific law is formulated may, and usually does, contain both general terms, in which no reference is made to any frame of reference, and particular terms, which do refer to some particular frame. Examples of general terms are: "electron," "massive particle," "coincidence." The simplest particular terms are the components of the coordinate quadruple themselves, x, y, z, and t. These coordinates may be taken to describe the results of certain manipulations with the meter sticks and the clocks of the system S. The algorithms in most common practical use contain general terms and in addition particular terms pertaining to only a single frame of reference, perhaps the frame provided by our laboratory.

Suppose now that we have a potential algorithm before us, formulated in general terms and particular terms pertaining only to S. Then the first postulate states that this algorithm will not enable us to calculate *correctly* the develment of the chain of events to which it ostensibly applies — that is, the algorithm will not correspond to a "law of nature" — unless it also enables us to calculate correctly the chain of events when we replace every particular term referring to S by the corresponding term referring to S'. In particular, this means that, given the algorithm A valid in S, in order to formulate the algorithm A' valid in S', we must *at least* replace x, y, z, t wherever they occur explicitly in A by x', y', z', t'. This may not be sufficient, because A may contain terms in which x, y, z, t are contained only implicitly so that they do not proclaim themselves. Notice that the converse of this state of affairs by no means holds. An algorithm formulated symmetrically in terms of S and S' is not necessarily valid because of the mere fact of symmetry.

The validity of any algorithm is subject to the check of experiment.

The task of an adequate theory of relativity is to find the complete set of algorithms by which all the phenomena in its universe of discourse may be determined. The search for such algorithms is greatly facilitated by the knowledge that the algorithms successfully used in prerelativity physics are approximately valid, and become entirely valid in the limit when the relative velocity of S and S' approaches zero. Thus, although we know from our analysis above that conservation of Newtonian momentum cannot yield a correct description of the actual development in time of mechanical phenomena, we know that it gives an approximately correct description and that any needed modifications will be small. The modification is, in fact, as follows. It is not the Newtonian momentum mv which is conserved in the interaction of bodies on each other, but the "relativity momentum" $mv/\sqrt{1 - \beta^2}$, where β is the ratio v/c. The correction to the Newtonian conservation law thus becomes important only for velocities approaching the velocity of light.

The "m" which appears in the expression above for the relativity momentum is, like the "m" which appears in the Newtonian momentum, a parameter characteristic of the particular body, a tag which the body carries with it, the same in all reference systems. To emphasize this, a special notation m_0 is adopted in relativity formulations. This "m_0" is the "rest mass," the same in all frames. The relativity momentum is not the product of the rest mass and the velocity, but is $m_0 v/\sqrt{1 - \beta^2}$. If we want to call the factor by which we multiply velocity to obtain momentum the "moving" mass, then we have: "Moving mass" is equal to "rest mass" multiplied by $1/\sqrt{1 - \beta^2}$. The moving mass is not constant, but approaches infinity as its velocity approaches the velocity of light. A moving mass which is infinite in the system S is also infinite in the system S'. In this respect, moving mass is unlike the infinite velocity of a sweeping searchlight, which is not infinite in both frames. These last remarks apply to linear motions along the x-axis; the situation is more complicated for motions in other directions.

In our formulation of the conservation law just given, the quantities v and β are particular quantities, characteristic of the frame of reference, and m_0 is a general term, independent of the frame, characteristic of the body itself, and determinable, given only the body. In terms of these particular and general quantities, the law of the conservation of momentum takes the form: Given a system of bodies with different masses m_0 which are allowed to interact with each other, then it will be found that the sum of the quantities $m_0 v/\sqrt{1 - \beta^2}$, formed for all the bodies, will be the same before and after interaction, and it will also be found that the sum of the quantities $m_0 v'/\sqrt{1 - \beta'^2}$ will be the same before and after interaction.

Suppose that we have found some candidate for a "law of nature" by successful formulation of the behavior in some particular frame of reference. How shall we know whether some specific term in this formulation is to be taken as a "general" or as a "particular" term? In the Newtonian formulation the term "m" was taken as a general term. We had to modify it and replace it by $m_0/\sqrt{1 - \beta^2}$, where "m_0" is now treated as a general term. What is the general situation here? Suppose, for example, that we have found that the force exerted by a stretched cord is equal to the product of the amount of stretch and the elastic constant of the cord. Which of the terms here are special and which are general? It is evident that "stretch" is a particular term, because it is defined geometrically in terms of spatial coordinates, and as such comes within the province of the transformation equations. But what about "force" and "elastic constant"? Are these general or particular?

One might attempt to answer these questions by an extension of the analysis we applied above to the conservation of momentum. If one does, one speedily encounters complexities of almost baffling difficulty. It is true that there are certain simple things that one can see qualitatively. For example, the concept of "rigid body," which is fundamental to the Newtonian mechanics of extended bodies, has to be modified. Because meter sticks change length when set in motion, no body behaves "rigidly" when set in motion. The first postulate, therefore, demands that wherever it occurs in a prerelativity formulation of a law of nature, the phrase "rigid body" be replaced by some other phrase slightly different from it. One of the consequences of this is that terms such as "density" must also be modified before they can be incorporated in a valid law of nature. We can also see qualitatively that, if it is the relativity form of momentum which is conserved, simple accelerations are not going to play the fundamental role which they did formerly. It is not mass times acceleration which is significant, but time rate of change of momentum. Furthermore, if no body is "rigid," the significance of an "elastic constant" is going to be modified, and we suspect that the elastic constant of a cord which determines the force exerted by the cord when it is stretched is not a parameter of the material of the cord in and for itself, but is in some way entangled with any motion in which the cord partakes.

Altogether, the situation is too complex to be covered by any simple analysis, and it demands the use of the complex mathematical apparatus of four-dimensional vectors, which would take us beyond the scope of this "primer." We content ourselves here with the statement that in applying the first postulate in detail to a wide range of phenomena, many of the terms which prerelativity theory treated as "general," without reference to the coordinate frame, turn out to be "particular," in that they contain concealed reference to the frame. This particularity has to be made specific for the different types of phenomena in order to get a formulation valid for all frames. In general, the modifications thus demanded are small for velocities small compared with the velocity of light. In the actual working out of the details it appears that electrodynamic systems are simpler than ordinary mechanical systems. A reason for this is that one of the fundamental laws of prerelativity electrodynamics automatically satisfies the first postulate of relativity; this is the law of the conservation of electrical charge. Once one has command of the electrodynamical situation, the situation for mechanics can be worked out by way of the behavior of compound systems in which both electrodynamical and mechanical effects enter. Historically, this is the way it was done. The formulas in general become exceedingly complicated. Thus "energy" is not the simple scalar it is in Newtonian mechanics, but becomes a four-dimensional tensor with sixteen components. Involvements with elastic stress and the flow of momentum appear of which there was no former inkling.

It would probably be exceedingly difficult to formulate in logically satisfactory terms all that the relativist would like to include under "law of nature." It is probably true that when confronted with any particular case, it would not be difficult to obtain a tolerable consensus as to whether a "law" were involved. But any such consensus is evidence of the common experience of contemporary physicists rather than of anything more fundmental. There are paradoxes and inconsistencies in common usage. It bothers no one that the velocity of some particular cannon ball is different in one frame of reference from its velocity in another. It is obviously not a "law of nature" that some particular object has some particular numerical velocity, and in general, numerical values of velocity change from frame to frame. But it *is* a law of nature that the velocity of light has the particular numerical value which it has, the same in all reference frames. This applies not only to light. If the velocity with which a searchlight sweeps the sky is c in one frame, it is also c in other frames. Is it a "law of nature" which secures this invariance? We would not call it a manifestation of a law of nature that a certain gyroscope on this earth is moving toward the nebula in Andromeda with a velocity of, say, 100 km/sec. By the same token, we should not say that it is due to a law of nature that the axis of the gyroscope persists in pointing toward the same nebula. But I suspect that many physicists would want to say just this. It is not a "law of nature" that the entire cosmos is moving with respect to our frame of reference with some particular velocity. The velocity of the cosmos is not the same in S and S', but what statement could better qualify to serve as a law of nature than a statement about the cosmos?

"Laws of nature" play an even more important role in general relativity theory than in special theory. What is meant by a formulation of a law being the "same" in two reference systems is made more explicit. By the "same" in two reference systems, general theory means *symbolically* the same, in the sense that a single symbolic formulation holds for different frames. If application is made to any concrete case, the numbers corresponding to the symbols are, of course, different in different frames. In many expositions of the general theory, including even those of Einstein, great importance is attached to finding the form of the "laws of nature" which shall be co-variant in all systems of generalized coordinates (that is, which shall have the same symbolic form). The implication usually is that here we have a powerful tool to assure that we are dealing only with "laws of nature." Actually, this criterion yields nothing. Any sort of specific behavior whatever, such as conservation of momentum in the classical Newtonian form, can be expressed in co-variant form in generalized coordinates, provided only that we are willing to pay the price in complication. Einstein admitted this on one or two occasions, but he kept forgetting it.

In special theory, a necessary condition of a "law of nature" is that it should be symmetrically formulated in symbols in S and S', but this is by no means sufficient. What more is necessary usually has to be left to the good sense of the physicist. This requirement of symmetry is usually thought to be rigorously satisfied by the equations as usually formulated, but actually it is not. If S and S' were fully equivalent, then formally the Lorentz equations should be *completely* symmetrical in the unprimed and primed letters. Actually they are not, as inspection shows. Physically, the reason for this is that the systems have been so set up that S' is moving with velocity v with respect to S, but S is moving with velocity $-v$ with respect to S'. Complete formal symmetry would demand that both velocities be the same. This can be accomplished at once by reversing the sign of the x'-axis as it is usually drawn in S'. If in the formulas on page 7 we replace x' by $-x''$, complete symmetry will be attained. Everyone knows how to handle the asymmetry in the equations as ordinarily written, and it does no harm. It is, however, curious that apparently no one has commented on the fact that, formally, the Lorentz equations do not satisfy the first postulate of relativity. It is too late to try to change the equations now.

The Observer

Let us now turn back again to some of the simpler things which are conceptually at the foundation of the entire edifice. The transformation equations deal with the coordinates of events in one or another frame of reference. How are these coordinates obtained in any actual case? Nearly all elementary expositions speak of "observers" associated with the frames. The x, y, z, t coordinates of the S frame are determined by an "observer" situated in the S frame, and the corresponding x', y', z', t' coordinates of the S' frame are determined by another "observer" situated in the S' frame. These observers are, tacitly, replicas of ourselves. When Einstein says that "natural phenomena run their course with respect to S' according to exactly the same laws as with respect to S," he is thinking of observers in S and S' who express the "laws" of what they see happening in exactly the same words, and who describe in the same words the operations in terms of which they define phenomena. The observer in S says that he measures lengths by successive applications of the meter stick which is the basis of the coordinate frame S, and the observer in S' states that *he* measures lengths by successive applications of the meter stick which is the basis of his coordinate frame S'.

There can be no question but that this expository device of two "observers" has great advantages in making vivid the significance of many statements of relativity theory and in making them intuitively acceptable. One often hears talk of how this or that appears to one or the other observer, or what one or the other would say about the situation. However, this point of view is not always straightforward, and there are concealed pitfalls, as shown by recent papers in the literature (Roy Weinstein, *Amer. Jour. Phys.*, **28,** 607–10 [1960]; J. Terrell, *Phys. Rev.*, **116,** 1043 [1959]), where it appears that what the observer actually sees may be quite different from what he has uncritically been assumed to see in most of the popular expositions of relativity since the very beginning. In particular, the observer does not *see* the Lorentz contraction of a moving meter stick. The fact that such errors have persisted with no reaction on any of the results which have been validly deduced from the theory would strongly suggest that the device of an observer traveling with the frame of reference plays no essential part and could be dispensed with. Detailed consideration shows that this is indeed the case.

The function of the observer is to determine the coordinates, in the frame in which he is moving, of some system of events. These coordinates describe the coincidences of the events with the nodes of the framework; the observer observes these coincidences. But a coincidence of an event with a node in the frame S is also a coincidence with a node in the frame S', and since a coincidence is a coincidence, the observer is in a position to observe the S' coincidences as well as the S coincidences. That is, only one observer is necessary, not two. Or going still further, an observer residing in S is not necessary. A single observer, in neither frame, observing from a detached external position, could determine all the coincidences in both S and S' and thus obtain the coordinate quadruples in both S and S' which are needed for the formulation of the results of the theory. Furthermore, the observer may be *any* observer. He might be you or he might be I. In our thinking, doubtless the observer in terms of whom you tacitly think will be you, and the observer in terms of whom I tacitly think will be I, but this will be of no importance because, given the frames S and S', both you and I will come out with the same coincidences and the same numbers associated with them. In this sense the events you and I observe are "absolutes." The interchangeability and one-to-one correspondence we have here between the private reports, yours and mine, are all that is necessarily involved when we demand that science be "public." To think that anything more is involved in any publicity of science is sheer metaphysics in its bad sense.

It is nevertheless undeniable that in the thinking of many physicists the device of an observer riding with the frame of reference plays a very real role in making vivid the content of the theory and in giving an intuitive feeling of understanding. From this standpoint, the meaning of such an expression as "the laws of nature in the frame S" becomes transparent and the fundamental postulates of the theory acquire a certain naturalness and even inevitability. But just because of this I think there is very real danger that this method of approach will give a feeling of false security. The danger is perhaps greater in the context of the general theory than in the context of the special theory. In the general theory we have various "cosmological" principles, to the effect that the universe always presents the same appearance to an observer, no matter how unusual the circumstances of this observer, whether situated at remote points in space or remote epochs in time. Now what should be the strongest motivation impelling a search for either a special or general theory is the realization that nature cannot be extrapolated — that under new ranges of conditions nature may behave in new ways. The attempt to command these new ranges by imagining an observer functioning like the observers of everyday experience would seem to be begging the question. The crucial conditions of cosmology, the remote reaches of time and space and the intense gravitational fields, which are the natural province of general relativity, are obviously beyond everyday experience. The same sort of remark applies, although in less degree, to special relativity, in which we are concerned with velocities of the order of magnitude of the velocity of light, which are beyond most everyday experience.

All things considered, I believe it is seriously to be questioned whether the device of different observers, each attached to his own frame of reference, has not been of greater disservice than service in promoting real understanding. In addition to the disservices implied in the last paragraph, the device of multiple observers tends to make one forget the existence of important limitations inherent in the use of even a single frame of reference.

Single Observer

Accepting now the possibility of getting along with a single detached observer, what does the special theory mean to him? In the first place, he sees before him two frames of reference S and S' moving with respect to each other. These are physical frames, constituted of a scaffolding of meter sticks with clocks at the nodes. Furthermore, they are identical frames in the sense that when they were originally stationary with respect to each other, or when in the future they may again be brought to rest with respect to each other, no detectible difference can be found between them (except for clock *settings* and displacements of origin). There is no reason why we may not suppose the observer himself to have constructed these frames and to have set them into relative motion. In doing this he will have set the clocks by a procedure we have already analyzed in detail. We suppose, furthermore, that the observer has been able to select his frames so that they are Galilean frames. Having now the frames, the observer makes various physical "measurements." In general, a physical measurement involves a physical instrument of some sort, the precise construction of which will depend on the sort of thing we are interested in. Whatever the type of measurement contemplated, we suppose that each frame is equipped with its own appropriate instrument. The instruments in S and S' are identical in the sense that, when both are brought to relative rest in the same place, no essential difference can be detected between them.

One of the simplest sorts of measurement the observer can make is of the relative velocity of S and S'. He observes the relative velocity of S' in S by combining in a specified way the x, y, z coordinates with the t readings on the clocks in S as S' sweeps past it, and he obtains the velocity of S in S' by combining in the same way the corresponding prime coordinates. He finds that one relative velocity is the negative of the other. A still simpler sort of measurement is of length. The instrument here is so simple that we are not likely to think of it as an instrument at all. There is nevertheless an instrument, the meter stick, which happens to have been built into the frame of reference. A great many measurements can be made by a proper combination of meter sticks and clocks — in fact, all the measurements of kinetics. If the observer is required to cover a broader range of phenomena than can be characterized by a simple kinetic description, he will need more complicated instruments than the built-in meter sticks and clocks, perhaps galvanometers. If he needs a galvanometer, he will have to provide himself with two, one stationary in S and another, of identical construction, stationary in S'. The pointers of the galvanometers provide new coincidences which have associated coordinate quadruples in S and S'.

In general, the raw material collected by the observer, the material about which relativity theory makes pronouncements, is sets of readings with *two* sets of instruments, one moving with respect to the other. This is a more faithful description of what the observer, you or I in our laboratory, actually does than to say that the observer employs two frames of reference. The reports of an experiment are almost always given in terms of a single observer, not two. Michelson did not report his interferometer experiments in terms of two observers, one, Michelson_1, observing at 6:00 A.M. and another, Michelson_2, observing at 6:00 P.M. The one set of instruments gives the unprimed coordinate quadruples in S and the other set of instruments gives the primed coordinate quadruples in S'. Relativity theory makes statements about the way in which measurements made with these two sets of relatively moving instruments are related to each other; it says that the corresponding coordinate quadruples will be correlated by the Lorentz equations.

In practical tests of the correctness of relativity theory, the conditions are not usually as clean-cut and as definitely articulated as in the ideal analysis above. We do not usually have two complete frames of reference or even two sets of instruments simultaneously observed, but we are more likely to have a single frame of reference and a single instrument, observed at different times and under different circumstances. This is obviously the case in the Michelson-Morley experiment, for example. The rigid arms of the interferometer provide the space framework, and the position of the interference fringes provides indirectly the equivalent of a clock. With this single space framework and clock we make observations under different circumstances — by rotating the whole interferometer we observe the effect of the orientation of the frame, and by observing at twelve-hour intervals we observe the effect of changing the velocity of the frame by a known amount in virtue of the rotation of the earth. A procedure like this has been easily accepted by the physicist as equivalent to the idealized procedure with two frames and two sets of instruments, although it would doubtless be a matter of great difficulty to show the strict equivalence in logic or to devise a set of axioms that would faithfully reproduce all that is involved.

A lively realization that relativity theory is essentially concerned with measurements made with *two sets of instruments*, one moving with respect to the other, I believe will go far toward relieving the feeling of malaise and intuitive revulsion that so many feel when first confronted with the theory. Consider, for example, the pronouncement that the velocity of light is the same in all reference systems. This is repulsive to common sense, for we know that when we change the velocity of the object with respect to which we are determining velocity, we change the velocity of every other object with respect to it. And what can we mean when we talk about the velocity of light *in* a frame of reference other than the velocity with respect to that frame? But our feelings change when we realize that two sets of instruments are concerned. Might it not be that when we change the instruments we compensate in whole or in part for the change we introduced by changing the body of reference? We can accept such a possibility without intellectual outrage. In fact, detailed consideration of what is involved when we measure the velocity of light in the moving system shows that exactly this compensation has taken place. This compensation becomes particularly obvious if we construct our clock according to the only specifications we have been able to formulate explicitly. This clock is a meter stick, with mirrors at the two ends between which light bounces back and forth, and in which time is measured by counting the number of passages of the light. The clock in the moving system is this arrangement transported. The postulate that the velocity of light is independent of the motion of the system now becomes the most banal truism if we reflect that we might make the beam of light whose velocity we are measuring the beam which is reflected back and forth between the mirrors of the clock. In fact, the truistic element in this situation is so overwhelming that we wonder whether we have said anything with any physical content at all when we say that light velocity is the same in all frames of reference. If the only possible definition for a clock were a meter stick with mirrors, then we would indeed have said nothing with any physical content. But the physical content gets in indirectly by way of the contention that clocks may be defined in other ways through the laws of mechanics instead of the laws of light propagation, and that the two definitions are consistent. However, the specific definition in mechanical terms seems still to be given, and from a purely logical point of view the situation would appear not to be entirely satisfactory even yet.

Velocity "Here and Now"

In the last paragraph we were concerned with measuring the velocity of light. The method of measurement there adopted was by incorporation of the beam of light into the fabric of the clock with which velocities are measured. Such a measurement ascribes physical meaning to the velocity of light "here and now." In general, it would seem that if "velocity of light" is to have physical significance, it must have something to do with where the light is — that is, with the here and now. However, all methods of measuring the velocity of light do not have the connotation of the here and now. In particular, when the velocity is measured by measuring how far the light has traveled in a known time from its distant place of origin, the emphasis is not placed exclusively on what is happening here and now; it involves also an episode in the historical past which has no immediately understandable or significant connection with the here and now, particularly if there is no ether and if space is truly empty. The fact that there is a mathematical connection with the abandoned place of origin must be physically adventitious, an artifact connected in some way with the fact that we are dealing with Galilean systems, in which the velocity of light is constant by postulate. In consequence, it seems to me that the method of deriving the Lorentz transformations used in many elementary expositions, namely by demanding that the expression $x^2 + y^2 + z^2 - c^2t^2$ be invariant, is a method giving less physical insight than the simpler method used on page 7, in which we demand that whenever dx/dt equals c, dx'/dt' also equals c. For the former method emphasizes the continuing significance of the historical point of origin of the light, whereas the latter deals with what is happening here and now.

Consistency with the present point of view would demand that we seek the factors which determine the future course of the light packet which is now passing us in what is happening here and now in the packet, not in what happened at the point of origin. That is, the point of view demanded is that of Huyghens' principle. It is perhaps significant that the name Huyghens does not occur in the index of names of either Von Laue's book or Pauli's, or in the popular books of Einstein.

The differential equations of the electromagnetic field ensure the operation of Huyghens' principle.

The Miracle of Radiation

From the macroscopic point of view, the packet of radiation, as it is made manifest by such macroscopic instruments as the motes of dust in its path, presents the same "miracle" as the massive particle which continues to move for all time in accordance with Newton's first law of motion. The "miracle" is that the future course of the packet, in both direction and velocity, is determined, whereas there does not seem to be enough feature in the here and now to effect such a determination. But the differential equation of the field supplies a fine structure, as it were, to the here and now, and a determination of the future course of the radiation in the context of the differential equation does not appear so utterly baffling. With this insight, one is tempted to go back to the massive particle and the incredible first law of motion. Perhaps there is also some sort of "fine structure" back of the particle to keep it on its course. One can dimly descry a clue here which might have led to the development of wave mechanics.

It is to be noticed that radiation has two stages of fine structure beyond the macroscopic packets we experience with our senses. There is the fine structure of the Maxwell equations, and beyond that the individual photons. Conventional relativity theory is not concerned with the second step toward the microscopic — that is, it does not concern itself with photons. Something similar doubtless holds for "matter waves," and these waves must have their photons.

We have been considering the packet of radiation, here and now, as a self-existent thing, divorced from contact with its origin. Does this mean that the slug does not carry with it any earmark impressed by its origin? The Doppler effect at once comes to mind. The Doppler effect is usually expressed in terms of an alteration of frequency, or difference of two frequencies, and this alteration of frequency is usually described from various points of view. Given a stationary frame of reference S, we may have a source either stationary or moving in S, and we may have an "observer" either stationary or moving in S. The original frequency may be either the frequency of the source in the frame S in case it is stationary; or, if the source is moving, the frequency of the source in a frame S' in which it is stationary. The received frequency, if the observer is stationary in S, is the frequency observed with local fixed clocks in S; or if the observer is moving, it is the frequency on local clocks fixed in a system S'' in which the observer is stationary. In all, one has to consider four altered frequencies corresponding to the four pairings of stationary or moving source with stationary or moving observer. The complexity of this situation is cut in half if, instead of considering a packet of radiation as permanently identified with its source, one considers the packet of radiation as an independent entity, here and now. Such a packet has associated with it a parameter, its length, which is characteristic of it and which it carries with it. This length can be traced back to the source, where it was impressed, so that there is in fact a connection with the source; but for the future course of the packet this early association with the source becomes irrelevant. The length of the packet, as it passes any fixed point, can be measured in terms of purely local operations. This length can also be measured in the frame S' of the Lorentz transformation. Application of the equations shows at once that the lengths in the two frames are connected by the relation:

$$\lambda' = \lambda \sqrt{\frac{1 + \frac{v}{c}}{1 - \frac{v}{c}}}.$$

This formula holds for the distance between any identifiable features in the stream of radiation of such a character that a length can be associated with it. In particular, if the source is emitting pure monchromatic radiation, there is a wave-length, the distance between successive nodes where both E and H are zero, and the above formula applies to this wavelength as measured in S and S'. There is an associated frequency, the frequency with which the nodes pass a stationary point. Since the velocity of propagation is the same in S and S', this frequency is inversely as the wave length, or:

$$\nu = \nu \sqrt{\frac{1 - \frac{v}{c}}{1 + \frac{v}{c}}}.$$

These formulas contain the conventional formulas for the first-order Doppler effect when the source and receiver preserve a constant relative velocity during the time of passage of light from one to the other, a condition which can be formulated only in the context of a frame of reference. It is only in this context that the question arises as to the symmetry of the Doppler effect in the velocities of source and receiver. A recent analysis has been given by Dingle of some aspects of the Doppler effect and certain paradoxes have been emphasized as compared with the usual understanding of the situation. These paradoxes do not arise in the context of the analysis above. In general, it would seem that an analysis in terms of the self-contained properties of the packet of radiation gets closer to the physics of the situation and is to be preferred to an analysis from the point of view of Dingle.

Summary and Discussion

In this final chapter I shall try to gather together the main threads running through the argument.

We have tried to find a minimum point of view which would read into the theory no more than is necessarily implied in the various successful applications of the theory to concrete situations. We have accepted the theory as giving correct experimental results and to that extent not needing criticism or revision. But various conceptual attitudes toward the theory are possible without involving experimental inconsistencies. We have been concerned here to find the conceptual attitude which would demand the least commitment on our part. The search for a minimum point of view is handicapped at the very start by the implications in the accepted name of the theory itself— the theory of *relativity*. There is supposed to be a deep philosophical principle back of this, to the effect that "everything is relative" and, in particular, is relative to the observer and his frame of reference. Superficial justification of such an attitude may be offered by such easy observations as that "position" means position of something with respect to something else, and "velocity" means velocity of something with respect to something else.

The name "relativity" and the attitude it naturally engenders are unfortunate, because they make more difficult the realization of the fundamental fact that there is an "absolute" background implicit in the theory without which its very language is meaningless. For beyond all frames of reference and their correlated observers, the theory assumes a system of "events," existing in and for itself, independent of the coordinates that describe it in any particular frame, the "same" for all observers and all frames. The task of every individual physicist is in some way to make himself master of the background system of events. This mastery involves in the first place exhaustive description, and after that the acquiring of understanding. The individual physicist will make unrestricted use of all the tools on which he can lay his hands. Among these tools is the so-called "frame of reference," and the physicist will not hesitate to employ one or several frames of reference as may suit his purposes. It is a fundamental thesis of this exposition that instead of thinking of relativity theory in terms of two or more observers, each associated with his own frame of reference, it is sufficient to think in terms of a single observer who formulates his results in terms of one or another frame of reference. This single observer may be any observer — either you or I — and he formulates his results in a public language such that you and I can repeat and check for ourselves what the other may tell us.

The results you or I publicize are, in general, the readings of instruments, understanding "instrument" in a sense broad enough to include our sense organs. The "frame of reference" is itself a special sort of complex instrument, having built into it meter sticks and clocks. When I describe a situation in two frames of reference, I am essentially describing it in terms of measurements with two sets of instruments. In general, one set of instruments is associated with the one frame and the other with the other. Under the circumstances to which the special theory applies, one of these sets of instruments is moving with uniform velocity with respect to the other. This suggests the following statement in minimum terms of the partial scope of special relativity theory: Special relativity theory is concerned with correlating the measurements which a single observer makes on given physical systems, first with one set of instruments and then with another set moving with uniform velocity with respect to the first. In the body of the text we have tried to show that this corresponds to what the physicist actually does in such situations as presented by the Michelson-Morley experiment, and that it is a more faithful reproduction of what happens than a formulation in terms of different observers and frames of reference.

Formally, the theory states that the coordinates of events determined with two sets of instruments, one moving uniformly with respect to the other, are related according to the Lorentz equations. In applying the equations to any concrete system of events, two different types of situation present themselves which distinguish the approach and general attitude of relativity theory from earlier physics. In the first place, there is the realization that it is important and necessary to specify various details in the manipulation of the two sets of instruments with a nicety which was not formerly appreciated. In fact, detailed analysis brings to light concealed niceties in instrumental manipulation the mere existence of which was not formerly suspected. An example is the analysis by which Einstein showed what is involved in a judgment of simultaneity in different frames of reference. In the second place, relativity theory demands the existence of certain new physical effects which were not formerly suspected. An example is the impossibility of imparting a velocity greater than the velocity of light to any material object.

Relativity theory, as ordinarily expounded, does not attempt an articulate or systematic examination of what it is necessary to specify with regard to the details of instrumental manipulation or what fundamental physical facts afford a sufficiently broad base, but is mostly content to deal with these questions by implication and only to the extent that failure to consider them might embarrass the working physicist. Thus no need is felt to specify the nature of a meter stick or a clock or the details of their method of transportation from one system to another. But it is assumed that the actual materials from which we construct meter sticks and clocks do in fact function to a sufficient degree of approximation as the equations assume, and that frames of reference exist which are Galilean and which may be found by trial and error.

The main purpose of this essay has been to close some of the logical gaps in the usual presentations of the theory, without, however, anything in mind as pretentious as a self-consistent scheme of postulates. We began our operational analysis of the details of our instrumental manipulations by considering only a single set of instruments in a single frame of reference. We were, in the first place, able to give explicit directions for setting up a Galilean frame of reference instead of leaving it with the vague statement that Galilean frames of reference "exist." The explicit exhibition of a Galilean frame of reference throws a new light on the significance of the Mach principle, showing that the inertial behavior of local bodies does not demand the esoteric participation of all the distant masses in the universe and that the persistence of the axis of a gyrocompass in pointing to some distant star is only an artifact.

Within a single frame, perhaps the most important question that arises when we carry our operational analysis to the level demanded by relativity phenomena is how we shall assign a time to distant events. This is essentially the question of setting distant clocks, or "spreading time through space." The answer we give in practice is controlled almost entirely by considerations of convenience — there is no physical compulsion to adopt one method in preference to another. Einstein adopted one particular method of setting distant clocks, which, because it was in his choice, amounted to a particular "definition" of the simultaneity of distant events.

Going further, the concept of "velocity" as generally understood involves assigning a time to distant events and therefore depends on the way we set distant clocks or the way we spread time over space. Several kinds of velocity are to be distinguished, including one-way two-clock velocity, one-way self-measured velocity, and round-trip one-clock velocity. Of the various sorts of velocity, it is only the one-clock velocities, including in particular round-trip velocity and self-measured velocity, which are physically significant in the sense that they do not depend on the arbitrary way in which time has been spread through space. This means that one-way two-clock velocities correspond to no physical "reality" and that a determination of them is irrelevant in describing the physical nature of a system. Questions with real physical content, such as the isotropy of the propagation of light, have to be formulated in terms of one-clock velocities, not one-way two-clock velocities as usually implied. How the question of isotropy may be formulated is shown in detail in the text.

Reichenbach has made the most elaborate previous analysis of the question of acceptable methods of spreading time through space. His analysis was made with the assumption that there is an absolute significance in the time order of events causally connected, even though they may be separated in space, and that signaling with light affords the most rapid possible causal communication. In the text, I criticize the analysis of Reichenbach from several points of view. Causal connectedness of distant events is not necessarily connected with order in time, but may just as well be correlated with directional effects in space. There are methods of setting distant clocks which do not involve causal propagation; and, in particular, clocks may be set consistently by transport, contrary to Reichenbach's explicit statement, and with no direct involvement with the properties of light.

When dealing with two sets of instruments, questions arise which are not so pressing with respect to a single set. In the first place, it is necessary that the two sets of instruments — meter sticks, clocks, whole frame of reference — be identical, in the sense that when the two sets are brought to relative rest in the same place no difference between them can be established by any sort of "direct" physical test — that is, any test not involving past history.

The question of the nature of a meter stick forces itself upon us in the context of two sets of instruments. It is shown that what might be intuitively thought to be the simplest meter stick does not in fact function as one when set into motion. We had to leave the question in practically the same unsatisfactory state as it is left by conventional theory, namely with the statement that as a matter of fact actual materials function as meter sticks as specified by the equations. This demands that, when a meter stick is set in motion, interatomic forces are automatically called into play which produce the relativity shortening; but we were not able to specify further what the nature of these interatomic forces might be.

The question of the nature of a "clock" is fully as pressing in the context of two sets of instruments as that of the nature of a meter stick. It is usually assumed without further precise specification that radiating atoms function as clocks and continue to do so when set into motion, but there is no transparent reason why they should behave in this way. It is possible, however, to specify the detailed construction of one macroscopic clock, namely a meter stick with mirrors at the two ends between which light is reflected back and forth. If it is legitimate to specify a clock in these terms, and certain assumptions about the nature of light are concealed here, then the postulate of relativity that the velocity of light is the same in all frames becomes superfluous because this is secured by the construction of the clock. The relativity change of rate of the clock when it is set into motion is also secured by its construction. Be this as it may, there does not seem to be much connection between a clock so constructed and a radiating atom, and the fact that actual atoms as they radiate do function as clocks has to be left to naked postulate, much in the same way as the motional contraction of meter sticks.

Light plays a major role in relativity theory, and much of our exposition was devoted to analyzing the nature of this role. In the first place, the equations are not immediately concerned with light as such, but are concerned with the unique mathematical properties of a particular velocity c. Everything with velocity c behaves, according to the equations, in the same way as light. "Light" itself is an intuitive concept — conventional theory attempts to give specific instructions to ensure that one is certainly dealing with "light" no more than it particularizes with regard to meter sticks and clocks. It is assumed that we can know when we are dealing with light. Furthermore, it is assumed that this knowledge is in terms of things that happen here and now — it is not necessary to know where the light came from to know whether it is light or not. All its properties may be determined in terms of measurements here and now — in this respect light is like a "thing." Relativity theory is concerned almost exclusively with only one of the properties of light, its velocity, which can be determined by measurements here and now. It is found experimentally that this velocity is always the same, independent of the nature or velocity of the source. The postulate that the velocity of light is independent of the velocity of its source is indispensable to relativity theory and is a much more fundamental postulate than that of the equality of velocity in all frames of reference, which is automatically secured by the construction of the clock.

The Lorentz equations may be derived by demanding that the velocity here and now — that is, the velocity expressed in differential form dx/dt — is invariant. It seems better to derive the equations in this way than by demanding that the velocity expressed in finite form in terms of the distance traveled from a remote and historical point of origin be invariant, as in the usual deduction.

Relativity theory by itself can offer no insight into how it comes about that light has a characteristic velocity here and now, particularly when it is considered that any sort of medium with properties analogous to familiar material media is ruled out. This insight comes not from relativity theory, which is a macroscopic theory, but from the electromagnetic field equations of Maxwell. "Light" has a "fine structure" or microscopic aspect, not envisaged by relativity, in that it consists of fluctuating electric and magnetic fields tied together by the Maxwell equations. These fields are "real"; it makes sense to ask whether light is actually accompanied by an electric field, and it also makes sense to ask whether the field equations are "true." These questions can be answered by operations with physical instruments. If the light is of long wavelength, as in radio waves, the operations can actually be performed with instruments we now have; if the wavelength is shorter, we assume that the operations could be performed ideally. The field gives a physical background, analogous to the old ether, capable of determining the future course of the radiation which may be present here and now. The velocity of the radiation is determined through a constant in the equations of the field, and the geometrical course of the radiation by the operation of Huyghens' principle, which can be shown to be a consequence of the structure of the equations. The equations are meaningless, however, except in the context of a material frame of reference — identifiable points in space have to be assumed.

There is, then, a certain physical "reality" corresponding to radiation and to the equations which control it. This "reality" is the reality of instrumental readings obtained at points of otherwise empty space. The old-fashioned ether, with its identifiable points which were to serve as the anchorage for a material frame of reference, is gone, but the "reality" of the instrument readings remains. It is this instrumental reality and the corresponding field that constitute the transfigured and resurrected ether of the later writings of Einstein and Dirac. This new "ether" is essentially the aggregate of all possible instrumental readings at points not occupied by "matter." This is also essentially the "ether" of other recent writers such as Ives, who insists almost passionately on the necessity for an ether, and Builder, who demanded it in one of his last papers (G. Builder, *Australian Jour. Sci.*, **22,** 87–97 [1959]).

This line of thought need be carried only a little further to result in a major change of philosophical outlook. If one examines what the physicist actually does, or imagines what he might do, one sees that his raw material consists of readings with instruments and nothing else. To the extent that we can concentrate on the raw material, the external world *is* the deliverancies of our instruments. It makes no sense to ask in this context whether our instruments give us a "true" picture of the external world, and anyone who feels an impulse to ask this question has not got the point of view.

It is immaterial for the physicist whether the instrument readings are furnished by some instrument operating on "matter" or on "empty" space, provided only that he knows the manner of operation and the nature of the instrument. Some of our instruments report the properties of so-called "matter" and some of them properties of so-called "fields," categories which traditional philosophy has tried up to now to keep distinct. This dichotomy I believe must now be abandoned, for these things can be synthesized into a larger unity, the unity of the instrument. To the extent that it is desirable to maintain a distinction between the categories, despite the fact that it cannot be made sharp, "matter" would seem to be associated with the identifiable and permanent, and the "field" or the "ether" would seem to be associated with what is left from the synthesis of all possible instrumental operations. This applies only on the level of macroscopic experience. At the microscopic level, we recognize that the electron or other elementary particle is not identifiable, but a discussion of how best to categorize phenomena at this level would take us far beyond this discussion of relativity.

The aggregate of all possible instrument readings constitutes a sort of superreality, for which such categories as "matter" or "field" (or "ether") are more or less irrelevant. A partial anticipation of this point of view is contained in Hermann Weyl's *Was Ist Materie?* (p. 18): "I am firmly convinced that the concept of 'substance' has today exhausted its role in physics."

It is part of the ultimate task of the physicist to find what every conceivable physical instrument would report under every conceivable circumstance. It is also part of his task to process this raw material into as homogeneous a conceptual whole as he can. That is, it is part of his task to "understand." The processing of the raw material is performed by the human nervous system, and eventually the physicist must face the task of acquiring an adequate understanding of the functioning of his nervous system. For present purposes, however, we do not need to go as far as this, and there is plenty of room for thought in working out the implications of the realization that our raw material is exclusively instrumental.

The break with tradition comes in the realization that the part of our instrumental world associated with "empty" space has as much "reality" as the part associated with "matter." We are bathed in an ever present sea of radiation, from which there is no escape, but of which our unaided sense organs give us no inkling. Since we have no inkling, we have invented the concept of "empty" space, which we think of as somehow the background or container of all our experience. Perhaps the next generation, brought up from childhood with radio and television, will have the realization bred into its bones that space is never empty but is always full of an infinite variety of detail that can be conjured out of it by appropriate means. Without suitable instruments this detail eludes us, but with the instruments there is no limit to the wealth of detail awaiting discovery in the "empty" space around us. Recent spectacular discoveries in radio astronomy afford dramatic examples of this. The next generation may also come to feel that the operation of "isolation," which is so fundamental to all our present thinking, is an illegitimate operation. The present generation can perhaps learn to give lip service to the thesis that there is no such thing as isolation, but I think it will not be able to school itself out of a feeling of the miraculous when it contemplates two force-free massive bodies continuing to separate for all future time at a uniform relative velocity.

There is a superficial resemblance between our thesis that the world is instrumental and Eddington's contention that the world consists of pointer readings. Eddington's pointer readings are four-dimensional space-time coincidences, and his thesis is that the world can be eventually reduced to space-time coincidences. This thesis has always seemed to me suspect, for the coincidences have a *quality* as well as a mere existence. It makes a difference whether it is an electron or a proton that partakes in the coincidence that yields a particular pointer reading. Given only the space-time coincidences, the physical situation that gave rise to them can by no means be reproduced. Eddington's scheme is not sufficient for even adequate description, to say nothing of adequate theory. On the other hand, when we say that the raw material of the world is instrumental, the nature of the instrument is an essential part of the picture. We assume that the instrument is already given and known. When we use one sort of instrument, we say that it gives information about electrons, and when we use another kind of instrument, we say that it tells us about protons.

The physicist is not a unique human being; the raw material of the man in the street is instrumental just as much as the raw material of the physicist. The general philosophical outlook of all of us has somchow got to be reconciled to the realization that the raw material out of which we construct our world consists of instrument readings and *nothing else*. Our world *is* the result of the processing of this raw material by our mental machinery. This realization is something extraordinarily difficult for common sense to assimilate. Our language and our habits of thought are so constituted that we find it almost impossible to avoid talking about what is "there" in the absence of the instrument. We find it very congenial to say that the function of the instrument is merely to reveal to us a world independent of the instrument. The exigencies of daily life and social living together almost demand that we talk about the tree as continuing to exist whether we or anyone else is present to see it or not. Yet this manner of speech conceals an inner contradiction, which indeed has not escaped the notice of philosophers. For if the meaning of "tree" resides in a context of which the seeing instrument is a necessary part, how shall we talk without self-contradiction about the "tree" when the necessary part is absent? We can talk about topics like this only by the exercise of good will and in the context of our common social experience.

Even the ordinary common-sense point of view finds no difficulty in recognizing that the instrument may modify whatever it is that it detects, so that we may never have direct knowledge of things "in and for themselves." The thesis of quantum theory that the measuring instrument reacts on what it measures has not been nearly as difficult for common sense to accept as some of the tenets of relativity theory. What common sense finds difficult in quantum theory is that the reaction between measuring instrument and object of measurement is not itself determined but is governed by pure probability. But whatever the quantum theorist or the man in the street may really mean here, the difficulty remains of finding a logically coherent way of talking about the situation.

It is instructive that the physicist did not have to wait for quantum theory to encounter the logical difficulty of distinguishing between the instrument and the object of information. The difficulty was already present in connection with the electromagnetic field. The electric field at a point is defined in terms of the force exerted on an exploring test charge. (The fact that the actual definition is in terms of a limiting force as the charge is made indefinitely smaller is irrelevant, since the limiting process is specified merely to avoid complications arising from the reaction of the test charge back on the original distribution. For our purposes this complication is of no importance, and we may define the field strength at a point as the force experienced by unit charge when brought to the point in question.) The usual implication here is that the "field," something which exists in the absence of the test charge, is "measured" by the force on the charge, this force being a result of direct interaction between charge and field. However, a detailed working out of the mathematics shows that this is by no means the case. The test charge experiences a force because of the differential action of the Maxwell stresses acting on it, the stress on one side being greater than that on the opposite side. This inequality of the Maxwell stresses in different directions is a direct consequence of the *perturbation* introduced by the test charge into the originally homogeneous field. Without the test charge, the Maxwell stresses are uniform and there is no net force. The test charge does not respond to the field which was there before the charge was introduced, but responds only to the modifications in the field introduced by the charge itself. It is a fortunate result of the mathematics that the net modified Maxwell stresses give directly the original unmodified field. One might at first be inclined to regard this exact compensation of the mathematics as a bit of a miracle, until one makes the cynical observation that the Maxwell stresses were constructed with exactly this result in view.

"Field-in-the-absence-of-a-charge" is something of which we can have no direct instrumental knowledge, and it has no instrumental meaning. It is a purely constructional paper-and-pencil quantity, which we introduce to simplify our processing of the instrumental results. Furthermore, "field-in-the-absence-of-an-*instrument*" has no meaning, although I believe that most physicists are inclined to forget this and to "reify" the field. They reify the field when they think of the field as residing in some sort of medium, in which action is transmitted from one part to a part adjacent to it. There is no instrumental method of distinguishing between pure action at a distance and action through a medium.

I do not believe that a logically consistent scheme of talking about the total situation here is possible. A degree of consistency adequate for most purposes can be achieved by talking on different "levels." There is the level of the raw material — that is, the level of the deliverancies of our instruments. Beyond this there is the level at which we process the raw material. On this level we talk about "what" it is that the instruments "reveal" to us. The operations on this level are mostly pure paper-and-pencil operations. This level is to a large extent a construction of our nervous systems, and we shall eventually have to take more careful account of the nature of our nervous systems than we do at present. There is and there can be no sharp dividing line between the levels — for one thing, we have to find some way of specifying the construction of our instruments.

Two further remarks are necessary to clarify our contention that the raw material of our world is the total delivery of our instruments. In the first place, this by no means implies that we accept the positivistic contention that the world is nothing but a transformed system of sense impressions. The nature of our nervous systems which process the raw material is as essential a part of the resultant world as the raw material which is processed. At the level of our present discussion, the nature of the processing machinery has to be regarded as independent of the material it processes. In the second place, we are not compelled because of our thesis to adopt any dogmatic attitude with regard to a generalized causality principle. We recognize that it is possible to divide the deliverancies of our instruments roughly into a part arising from "matter" and a part arising from the "ether" or "empty space." Now there is a form of the generalized causality principle which maintains that it is possible to arrange into a logically coherent and causally consistent scheme that part of the total complex of instrument readings which pertains to matter only, provided that these readings are expanded to include similar readings through all past history, and are not merely the readings here and now everywhere at single instants of time. This generalized causality principle may or may not be true; it is for experiment to decide. There can be no question, however, that if we are going to have a causal "material" world, we shall have to include instrument readings at points of empty space if we restrict ourselves everywhere to readings at a single instant of time. Here again, as always, we find it impossible to maintain the logical sharpness we desire. What shall we mean by instrument readings "here and now" when we have to expand the instrument, as in the radio telescope, to cover acres of space? Furthermore, it is possible to push our analysis to a point where the "past" becomes only one aspect of the "here and now."

We have tried to understand the fact that light here and now has a characteristic velocity, in spite of the fact that there is no underlying medium in the old sense, by considering that light is an electromagnetic disturbance and as such satisfies the differential equations of the electromagnetic field. This "understanding" is adequate for many purposes, but there is no limit to the extent to which we can push our analysis, nor is there any authority to impose a moratorium on the questions we ask. If we ask why it is that the equations of the field are as they are, or seek to find some concealed mechanism which might afford an "explanation" of the equations, there is no present answer. From this point of view the equations merely describe what we find when we refine our instruments as far as we know how. Simple description, it seems to me, is the inevitable end of any analysis of which we are capable.

After our discussion of the properties of light, we asked whether relativity theory enabled us to acquire an understanding of the phenomena of mechanics comparable to such understanding as we have been able to acquire of the propagation of light. The fundamental problems are closely analogous. We formerly had to understand how light, here and now, in the absence of a medium, could have a characteristic velocity. We now have to understand how it is that a force-free massive particle which here and now has a certain definite velocity manages to maintain that velocity for all future time, cut off as it apparently is from all contact with its surroundings. Something analogous to the electromagnetic field equations, but applicable to inertial matter, seems demanded here, but there is nothing in sight that meets the requirements. In fact, the need for something analogous appears so imperative that one is strongly tempted to believe that there must be some new physical effect so small as to have hitherto escaped detection. It is well known that there is no Poynting vector for the flux of gravitational energy analogous to the Poynting vector for the flux of electromagnetic energy. The electromagnetic Poynting vector is proportional to the vector product of *E* and *H*, the electric and magnetic field strengths. Apart from magnetized matter, *H* originates in the motion of electrical charges. Under ordinary circumstances the *H* which thus arises is very small, and for a long time it escaped experimental detection although it was suspected and sought for. The electromagnetic field equations are relations between *E* and *H*. If there were such a thing as an "inertial *H*," we might have inertial field equations (the inertial *E* would be the ordinary gravitational force), and Newton's first law of motion would be no more miraculous or incredible than the constant velocity of light.

We finally have to try to understand how Einstein did it. There can be no question, I think, that his was not a minimum point of view. The device of two observers, each moving with his own frame of reference, plays too prominent a part in all Einstein's exposition to permit such an interpretation. The psychological attitude which finds congenial the device of independent observers, each associated with his own frame of reference, may be poles apart from the psychological attitude which prefers a formulation in terms of a single observer making his measurements with differently moving sets of instruments. It is perhaps for the psychologist rather than the physicist to attempt to recapture the full flavor of Einstein's intuitional drive, an intuition which felt complete confidence in its ability to put itself in the place of different observers in different circumstances, no matter how strange. This intuitional drive is even more manifest in Einstein's general theory of relativity than in his special theory. It is furthermore to be remarked that at the present time the ultimate acceptance of the general theory is in much greater doubt than the ultimate acceptance of the special theory. The latter would seem destined always to play a role in at least the same way that we believe Newtonian mechanics will always play a role.

There is a curious difference in the results of Einstein's intuitional approach when applied to the situations of the special and the general theory which has already been commented on. The special theory demands *numerical* invariance of the "laws of nature" in different frames, whereas the general theory demands only symbolic invariance. Numerical invariance has physical content; symbolic invariance has no physical content, something which Einstein knew with his intellect but which his intuition was continually forgetting.

As matters worked out, Einstein's intuitional approach yielded a powerful tool in those situations contemplated in the special theory. This approach was sufficiently different from that of his predecessors to enable him to complete an edifice which was rapidly approaching completion at the hands of Poincaré and Lorentz. In fact, Poincaré and Lorentz had so nearly completed the edifice that scholars like Ives and Whittaker have refused even to associate Einstein's name with the special theory. It would appear, however, from the writings of other authors, Pauli and Born in particular, that the position of Ives and Whittaker is extreme, and that Einstein did contribute something essentially new which made completion of the whole edifice possible. Einstein's new insight was that *all* the phenomena of physics are on the same footing — if electrodynamic mass is a function of velocity, as Lorentz and Poincaré knew it to be, then ordinary mechanical mass must be the same function. This is the sort of insight which it would seem would occur much more probably to a man with the general attitude of Einstein than to a man with the intuitional approach of Lorentz, for example.

In the confrontation of Einstein and Lorentz we have the confrontation of two diametrically different attitudes. Lorentz demands the detailed mechanism back of physical phenomena and does not feel that he has achieved physical understanding until he has found a mechanism. Lorentz follows in the great English tradition exemplified in high degree by Faraday, Maxwell, and Kelvin. One recalls the insistence of the youthful Maxwell, who demanded that he be shown "the *particular* go of it." The other attitude desires broad general principles and feels that it has achieved understanding when it can formulate the broad principles that apply. This attitude is pre-eminently exemplified by Einstein in his handling of general relativity theory. His need for understanding was satisfied with his discovery of the role of geodesics in general coordinates and his reduction of a gravitational field to a space-time curvature. The other field in physics dominated by this attitude is thermodynamics, in which we have phenomena controlled by broad principles, which come perilously close to being verbal principles in their lack of concern for any possible mechanism in the background. Thermodynamics is undeniably successful, and it has even often been claimed that its principles are the most general we have in physics. Nevertheless, most physicists find the approach of thermodynamics uncongenial and do not feel that they have achieved real understanding until they have reduced thermodynamics to statistical mechanics, despite the logical difficulties in this position, which I have pointed out in another place. It would seem that special relativity theory has still to find its statistical mechanics, although Fitzgerald and Lorentz went far along the road.

The very greatest physicists are capable of using either approach. Newton with his inverse square law of gravitation, which seemed to demand the intuitively repulsive action at a distance, was adopting the approach of general principles with no concern for any possible mechanism. It was in connection with his law of gravitation that Newton enunciated his often quoted "hypotheses non fingo," a dictum to be understood only in context. But Newton was as capable as anyone else of demanding and inventing special mechanisms in other contexts, as shown by his "fits of easy reflection" to explain interference phenomena, and by many of the speculations in his *Opticks*. In the same way, Einstein used the approach of general principles in his theories of relativity, special and general, but was equally capable of the detailed mechanistic approach, as shown by his theories of the Brownian motion and the photoelectric effect, which a number of qualified observers as late as 1922 regarded as of even greater importance than his theories of relativity.

We return now for a moment to the minimum point of view adopted in this exposition, the point of view that finds necessary only a single observer who makes use of one or another of different possible reference frames and who reports the results of the measurements which he makes with one or another set of instruments, one moving with respect to the other. It is my contention that this corresponds to what actually happens in the scientific enterprise. No scientific result can be expressed except in terms of the experiences, actual or potential, of individual scientists. There is no corporate or social observer — only single individual observers. But this single observer must be *any* observer, either you or I. So long as science is expressed in terms of procedures which either you or I could perform and the results of which you or I could verify, we have all that is necessary or indeed all that we know how to use. To the extent that science implies communication between individuals, either formal communications written in libraries for all to read or less formal and more ephemeral personal communications, it satisfies this requirement if the communication is in such a form as to be intelligible to any individual with adequate capacity and preparation. We do, as a matter of fact, make this requirement of any communication that we are willing to call scientific. In particular, special relativity theory, as expounded above in terms of the single individual observer, meets this minimum requirement, and I believe can be properly understood only from this point of view.

Index